A-Level Maths for OCR S1

Paul Sanders

Published in 2005 by:
Nelson Thornes Ltd
Delta Place
27 Bath Road
CHELTENHAM
GL53 7TH
United Kingdom

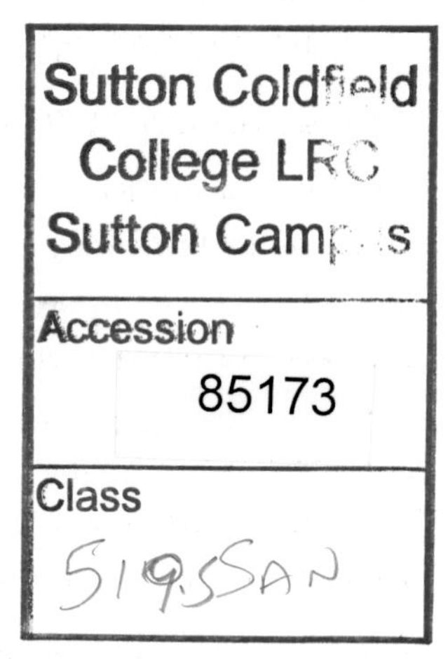

05 06 07 08 09 / 10 9 8 7 6 5 4 3 2 1

A catalogue record for this book is available from the British Library

ISBN 0 7487 9455 7

Sample paper written by Peter Balaam

Page make-up by Mathematical Composition Setters Ltd, Salisbury, United Kingdom

Printed and bound in Spain by Graphycem

Acknowledgements

We are grateful to Oxford Cambridge and RSA Examination Board for permission to reproduce all the questions marked OCR.
All answers provided for examination questions are the sole responsibility of the author.

The publishers have made every effort to contact copyright holders but apologise if any have been overlooked.

CONTENTS

INTRODUCTION

A-Level Maths for OCR is a brand new series from Nelson Thornes designed to give you the best chance of success in Advanced Level Maths. This book fully covers the OCR **S1** module specification.

In each chapter, you will find a number of key features:

- A beginning of chapter **OBJECTIVES** section, so you can see clearly what you should learn from each chapter
- **WORKED EXAMPLES** taking you through common questions, step by step
- Carefully graded **EXERCISES** to give you thorough practice in all concepts and skills
- Highlighted **KEY POINTS** to help you see at a glance what you need to know for the exam
- **EXTENSION** boxes with background information and additional theory
- An end-of-chapter **SUMMARY** to help with your revision
- An end-of-chapter **REVISION EXERCISE** so you can test your understanding of the chapter

At the end of the book, you will find a **MODULE REVISION EXERCISE** containing exam-type questions for the entire module. This is divided into four sections, mirroring the structure of the specification. [Each section tells you which chapters you should have done.]

Finally, there is a **SAMPLE EXAM PAPER** written by an OCR examiner which you can do under timed exam conditions to see just how well prepared you are for the real exam.

1 Presentation of Statistical Data: Diagrams and Averages

The purpose of this chapter is to enable you to

- use frequency bar charts, frequency polygons and vertical line diagrams to represent ungrouped data
- calculate means and medians of ungrouped data
- use histograms and cumulative frequency graphs to represent grouped data
- calculate estimates of means and medians

Types of Data

The data collected in a statistical survey are either **qualitative** or **quantitative**.

Qualitative Data

Qualitative data are non-numerical data, such as "Favourite flavour of ice cream" or "Colour of car". Analysis of qualitative data is usually restricted to diagrammatic representation in pie charts or bar charts, and the making of simple conclusions such as the most common car colour or the most common ice cream flavour.

The diagrams below show the types of medal won by the British and French Olympic teams at the 2004 Athens games.

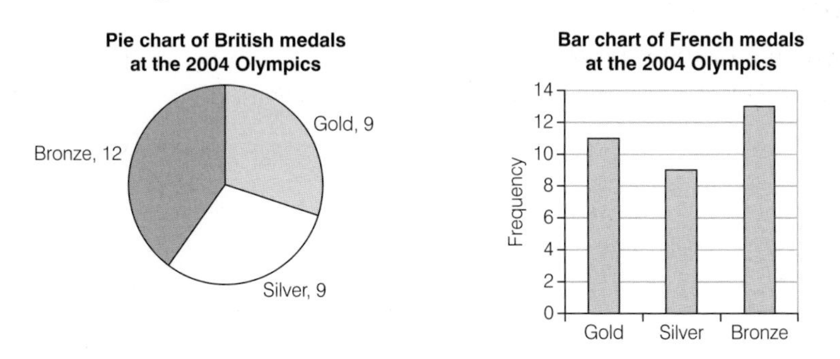

A bar chart is generally quicker to draw and easier to interpret than a pie chart.

Quantitative Data

Quantitative data are numerical data such as "Number of days absence during a term by students at a sixth form college" or "The weights, to the nearest kilogram, of a group of rugby players" or "The time spent travelling to work by a group of office workers".

Quantitative data will frequently be summarised in a frequency table. There are two types of frequency table: ungrouped and grouped. We shall see that the type of frequency table is a major factor in determining the appropriate way of analysing the data.

Working with Ungrouped Frequency Tables

Consider first the example of the number of days absence during a term of a group of 30 students at a sixth form college.

If the raw data is

4 1 2 1 2   4 5 5 2 4   4 6 5 3 5
7 5 2 1 0   3 5 4 4 1   3 3 3 6 5

then it can easily be summarised in an **ungrouped frequency table**:

Number of days absence	0	1	2	3	4	5	6	7
Tally	I	IIII	IIII	~~IIII~~	~~IIII~~ I	~~IIII~~ II	II	I
Frequency	1	4	4	5	6	7	2	1

Each frequency in the table relates to a **single** possible value that the data might take. Such a table can only be produced when there are definite distinct gaps between the possible values that the observations can take. We call such data **discrete data**.

An ungrouped frequency table is a completely accurate summary of the original data since the exact values of the original pieces of data can be recovered from the frequency table.

It is possible to deduce several important features of the data from the frequency table. For example:

- the most common number of days absence was 5;
- no student had more than 7 days absence;
- only one student had a 100% attendance record;
- 90% of the students had at most 5 days absence.

Diagrammatic Representation of Ungrouped Data

Discrete, ungrouped data can be represented diagrammatically by a frequency bar chart, a vertical line diagram or a frequency polygon.

EXAMPLE 1

Draw a suitable diagram to illustrate the data in the table below, which gives the number of days absence that students had during a term.

Number of days absence	0	1	2	3	4	5	6	7
Frequency	1	4	4	5	6	7	2	1

SOLUTION

Any one of these three diagrams would suffice:

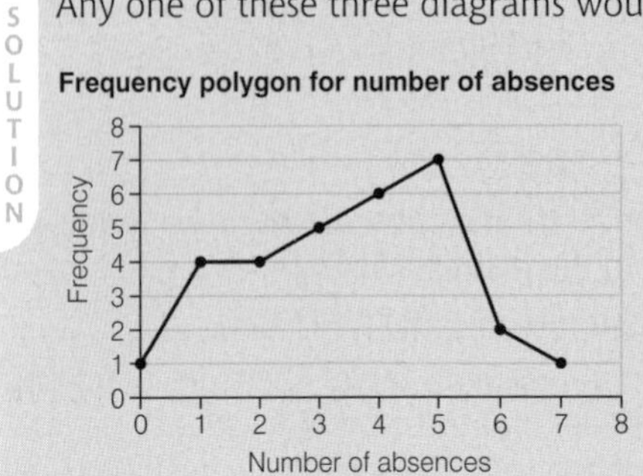

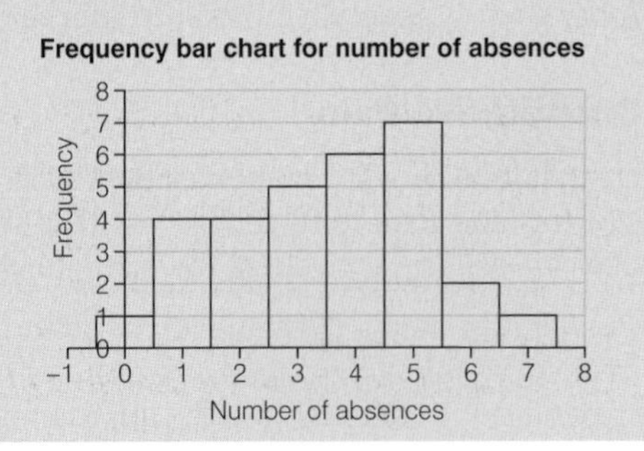

EXAMPLE 1 (continued)

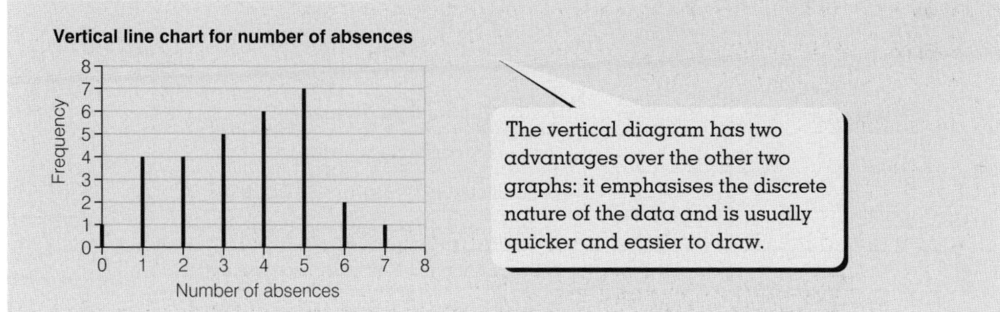

Averages

An average is an attempt to describe the typical member of the distribution. **An average is called a measure of central tendency**. There are three commonly used averages: the mode, the median and the mean.

The **MODE** is the observation that occurs most often. For the absentee data of Example 1 above, the mode is 5 absences since this had the highest frequency.

The **MEDIAN** is the middle observation when the data are arranged in order of magnitude.

EXAMPLE 2

Find the median of the observations 2, 7, 3, 7, 5, 8, 2

SOLUTION

When arranged in order, the data are 2, 2, 3, **5**, 7, 7, 8

so the median is 5.

When there is an even number of pieces of data, there is not a unique middle piece of data. In this case, the median is found by arranging the data in order, then adding the values of the two middle pieces of data and finally dividing the result by 2.

EXAMPLE 3

Find the median of the observations 3, 7, 5, 8, 2, 7

SOLUTION

When ordered the data are 2, 3, **5**, **7**, 7, 8

so the median is $\frac{5+7}{2} = 6$.

In general, if we have n observations arranged in size order then:

- if n is odd the median is the $\frac{1}{2}(n+1)$th observation;
- if n is even the median is found by adding together the $\frac{1}{2}n$th observation and the $(\frac{1}{2}n+1)$th observation and dividing the result by 2.

For larger amounts of ungrouped discrete data, the calculation of the median is best done with the aid of a **cumulative frequency table**.

EXAMPLE 4

Find the median number of days absence of the students in Example 1.

SOLUTION

Number of days absence	Frequency	Cumulative frequency
0	1	**1** student had 0 days absence
1	4	**5** students had 1 day or less absence
2	4	**9** students had 2 days or less absence
3	5	**14** students had 3 days or less absence
4	6	**20** students had 4 days or less absence
5	7	**27** students had 5 days or less absence
6	2	**29** students had 6 days or less absence
7	1	**30** students had 7 days or less absence

$1 + 4 = 5$

$5 + 4 = 9$

$9 + 5 = 14$

There are 30 pieces of data altogether, so the median is found by ordering the data and then adding together the 15th and 16th pieces of data and dividing the result by 2.

From the table we see that just 14 students had 3 or less days absence, but 20 students had 4 or less days absence. The 15th and 16th students therefore both had 4 days absence.

Median = $\frac{4+4}{2} = 4$ days absence.

The **MEAN** is usually denoted by $\bar{x}$, where $\bar{x} = \dfrac{\text{total of all the observations}}{\text{number of observations}}$.

EXAMPLE 5

Find the mean of the observations 2, 7, 3, 7, 5, 8, 2

SOL

$$\bar{x} = \frac{2+7+3+7+5+8+2}{7} = \frac{34}{7} = 4.86 \text{ (2 d.p.)}$$

The mean is by far the most important of the three averages. Several factors contribute to its importance. Each observation in the sample contributes to the calculation of the mean in the same way. The mean also has by far the easiest algebraic formulation, and this allows both easy computer or calculator calculation and simple deduction of fundamental properties. To understand the algebraic formulation for the mean you must be familiar with sigma notation, which you meet formally in the C2 module. The boxed text below summarises the important features.

Sigma Notation

In many branches of Mathematics we will often be interested in the **sum** of a whole series of terms, and it is useful to have a shorthand for such expressions so that we don't have to write out every term in the sum.

The shorthand used employs the symbol Σ, which is the Greek capital letter S known as sigma.

We write Σ as shorthand for sum.

The expression $\Sigma_{r=2}^{7} r^2$ reads as "the sum of r^2 for integer values of r between 2 and 7", so we can write

$$\sum_{r=2}^{7} r^2 = 2^2 + 3^2 + 4^2 + 5^2 + 6^2 + 7^2$$

Similarly

$$\sum_{r=1}^{4} 5 \times 1.2^r \text{ is shorthand for } 5 \times 1.2^1 + 5 \times 1.2^2 + 5 \times 1.2^3 + 5 \times 1.2^4$$

and

$$\sum_{r=1}^{100} \frac{1}{r^2} \text{ is shorthand for } \frac{1}{1^2} + \frac{1}{2^2} + \cdots + \frac{1}{100^2}$$

Your graphical calculator will probably allow you to use sigma notation to evaluate sums.

If we have a sequence of terms $a_1, a_2, a_3, \ldots, a_n$ then

$$\sum_{r=1}^{n} a_r \text{ is shorthand for } a_1 + a_2 + a_3 + \cdots + a_n$$

and

$$\sum_{r=1}^{n} r^2 a_r \text{ is shorthand for } 1^2 \times a_1 + 2^2 \times a_2 + 3^2 \times a_3 + \cdots + n^2 \times a_n$$

Properties of Sigma Notation

There are three simple properties of sigma notation that we will use frequently:

$$\begin{aligned}\sum_{r=1}^{n} (a_r + b_r) &= (a_1 + b_1) + (a_2 + b_2) + (a_3 + b_3) + \cdots + (a_n + b_n) \\ &= (a_1 + a_2 + a_3 + \cdots + a_n) + (b_1 + b_2 + b_3 + \cdots + b_n) \\ &= \sum_{r=1}^{n} a_r + \sum_{r=1}^{n} b_r\end{aligned}$$

If c is a constant, then

$$\sum_{r=1}^{n} (ca_r) = ca_1 + ca_2 + ca_3 + \cdots + ca_n = c(a_1 + a_2 + a_3 + \cdots + a_n) = c\sum_{r=1}^{n} a_r$$

and

$$\sum_{r=1}^{n} c = c + c + c + \cdots + c = nc$$

For example,

$$\sum_{r=1}^{4} 6 = 6 + 6 + 6 + 6 = 4 \times 6 = 24$$

When calculating the mean of observations $x_1, x_2, x_3, \ldots, x_n$ each occurring with frequency 1, we can write $\sum_{i=1}^{n} x_i$ or, more simply, Σx to denote the sum, or total, of the observations.

So $\bar{x} = \dfrac{\text{total of all the observations}}{\text{number of observations}} = \dfrac{\Sigma x}{n}$.

The mean of observations $x_1, x_2, x_3, \ldots, x_n$ each occurring with frequency 1 is given by

$$\bar{x} = \frac{\Sigma x}{n}$$

The calculation of the mean of an ungrouped frequency distribution is often presented in tabular form.

EXAMPLE 6

Calculate the mean number of absences of the students of Example 1.

SOLUTION

Number of days absence x	Frequency f	xf
0	1	0
1	4	4
2	4	8
3	5	15
4	6	24
5	7	35
6	2	12
7	1	7
	30	105

These 4 students had a total of $2 \times 4 = 8$ days absence.

Finding the total of the frequencies gives the total number of observations.

Finding the total of this column gives the total number of absences of all of the students.

Mean $= \bar{x} = \dfrac{\text{total of all the observations}}{\text{number of observations}} = \dfrac{105}{30} = 3.5$ days absence.

In general, if we have observations $x_1, x_2, x_3, \ldots, x_n$ occurring with frequencies $f_1, f_2, f_3, \ldots, f_n$, respectively, then

- Σxf denotes the sum of products of each observation by its frequency and gives the total of all the observations
- Σf gives the number of observations

so we can write

$$\bar{x} = \frac{\text{total of all the observations}}{\text{number of observations}} = \frac{\Sigma xf}{\Sigma f}.$$

The mean of observations $x_1, x_2, x_3, \ldots, x_n$ each occurring with frequency $f_1, f_2, f_3, \ldots, f_n$ is given by

$$\bar{x} = \frac{\sum xf}{\sum f}$$

The total number of observations, $\sum f$, is sometimes written as N so the formula for the mean can be written as $\bar{x} = \frac{\sum xf}{N}$.

Your calculator will almost certainly be able to calculate the mean of a set of data and you should ensure that you know how to enter a frequency distribution onto your calculator and how to obtain the value of the mean.

The next example suggests the amount of working you should show when using your calculator to obtain a mean.

EXAMPLE 7

The marks (out of 20) obtained by a group of students for an assignment are recorded in the table below. Calculate the mean of these marks.

Mark x	11	12	13	14	15	16	17	18	19
Frequency	1	0	2	5	7	8	5	2	1

SOLUTION

$$\bar{x} = \frac{\sum xf}{N} = \frac{480}{31} = 15.48 \quad \text{(2 d.p.)}$$

These values can be obtained from your calculator once you have inputted the frequency distribution.

EXERCISE 1

1 One week, a barber kept a record of the number of haircuts he completed each hour.

No. of haircuts	0	1	2	3	4	5	6
Frequency	4	8	12	9	7	4	1

a) Represent this frequency distribution with a suitable diagram.
b) Calculate the mean and median of the number of haircuts completed in an hour.

2 The table below shows the number of peas in pods of a particular variety.

No. of peas in pod	4	5	6	7	8	9
Frequency	40	90	110	95	55	20

a) Draw a suitable diagram to illustrate the data.
b) What is the modal number of peas in a pod?
c) Evaluate the mean and median number of peas in a pod.

3 The table below shows the marks obtained by a group of 60 pupils for some English homework.

Mark	1	2	3	4	5	6	7	8	9	10
Frequency	2	2	4	6	10	12	10	9	4	1

a) Draw a suitable diagram to illustrate these data.
b) Find the mean, median and mode of these pupils' marks.

4 An employer conducted a survey to investigate the number of spaces in the company car park at 10 : 00 each working day in March and each working day in August. The results are presented in the vertical line diagram shown below.

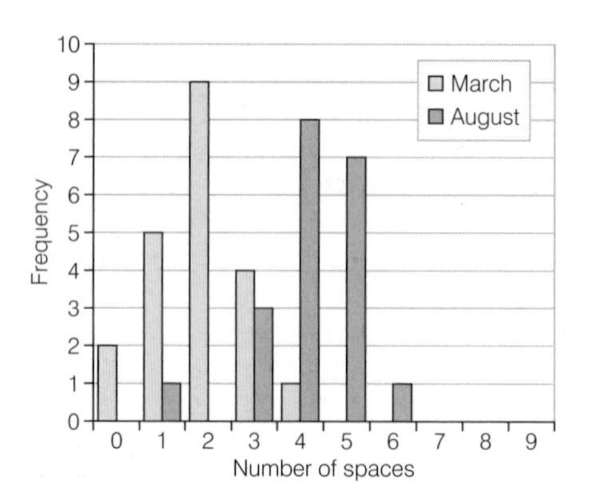

a) Calculate the mean and median of the number of spaces in the car park during March.

b) Calculate the mean and median of the number of spaces in the car park during August.

c) Write a short paragraph comparing the distribution of the number of empty spaces in the car park during March with the distribution of the number of empty spaces in the car park during August.

5 The route of the number 79 bus in a large city includes a stop at the city's main railway station. A bus stops there every ten minutes from 10 : 00 until 16 : 00. One day a survey was conducted to investigate the number of passengers getting onto the bus at the railway station. The data collected is shown below.

4, 9, 4, 12, 11　　8, 10, 14, 5, 6　　13, 9, 9, 3, 9　　3, 6, 4, 11, 6
9, 9, 7, 8, 6　　15, 9, 8, 11, 7　　7, 10, 10, 4, 4　　8, 9

Summarise the data in an ungrouped frequency table and hence determine the mean and median number of passengers getting onto the bus at the railway station.

6 **a)** A sample of 40 pieces of data gave $\sum x = 1268$. What was the mean value of the data?

b) A sample of data gave $\sum f = 80$, $\sum xf = 686$. What was the mean value of the data?

Working with Grouped Frequency Tables

In most surveys the number of different possible data values is so large that the production of an ungrouped frequency table is impractical because the table would be very large and probably meaningless. In such circumstances it is usual to collect sets of possible data values into groups and produce a frequency table for these groups. Such a table is called a **grouped frequency table**.

Consider the masses, to the nearest kilogram, of a group of 50 rugby players.

The raw data are:

106	97	104	101	110	109	121	97	70	117
99	122	120	106	108	76	96	111	103	83
114	95	128	101	99	99	132	109	81	104
123	96	125	89	119	96	118	87	107	108
100	104	124	90	99	111	100	120	104	107

The maximum mass is 132 kg and the minimum mass is 70 kg. There are 63 different integers between 70 and 132 (inclusive), and it is not appropriate to have a frequency table with 63 different values in it.

A possible **grouped frequency table** for the data is

When summarising data into a grouped table it is important that the groups chosen should be unambiguous: it must be clear that each piece of raw data fits into one and only one group.
The number of groups used is a matter of personal preference, but between 5 and 10 is usually sensible unless there is a large number of observations available, in which case more groups may be appropriate.
The groups can have different widths: since most of the masses are between 90 and 115 kg, relatively narrow groups are used between these limits and wider groups elsewhere.

Mass (kg)	Tally	Frequency
70–79	II	2
80–89	IIII	4
90–94	I	1
95–99	~~IIII~~ ~~IIII~~	10
100–104	~~IIII~~ IIII	9
105–109	~~IIII~~ III	8
110–114	IIII	4
115–124	~~IIII~~ IIII	9
125–134	III	3

Presentation and Analysis of Grouped Data

Grouped data is usually represented pictorially by either a histogram or a cumulative frequency graph.

Since the visual impact of a bar graph is made by the area rather than the height of the bars, grouped data is usually represented by a bar graph, called a **histogram**, where **the area of each bar corresponds to the frequency of the group**. The heights of the bars of a histogram are found by calculating the **frequency density** of each group using the formula

$$\text{Frequency density} = \frac{\text{Frequency}}{\text{Group width}}.$$

Histograms are particularly useful if a visual comparison of two distributions is required.

A **cumulative frequency graph** is usually drawn if the **median** of grouped data is to be estimated.

The **mean** of grouped data is easily estimated using a calculation based on the midmarks of each group as the x values. The result can only be an estimate of the mean since the exact values of the data have been lost in the process of producing the grouped frequency table.

EXAMPLE 8

The table below shows the masses, in grams, of 100 animals.

The usual convention is that the 50–70 group includes all the animals whose masses satisfy 50 ⩽ mass < 70

Mass	0–50	50–70	70–80	80–90	90–100	100–150
Frequency	9	15	18	26	21	11

a) Draw a histogram for the data and hence estimate the number of these animals with a mass between 53 and 78 grams.
b) Calculate an estimate of the mean mass of these animals.
c) Determine the median mass of the animals.

SOLUTION

a) To draw a histogram we must first calculate the frequency densities:

Mass	Frequency	Frequency density
0–50	9	0.18
50–70	15	0.75
70–80	18	1.80
80–90	26	2.6
90–100	21	2.1
100–150	11	0.22

The width of this group is 70 – 50 = 20.

The frequency density for this group is $\frac{15}{20} = 0.75$.

The width of this group is 100 – 90 = 10.

The frequency density for this group is $\frac{21}{10} = 2.1$.

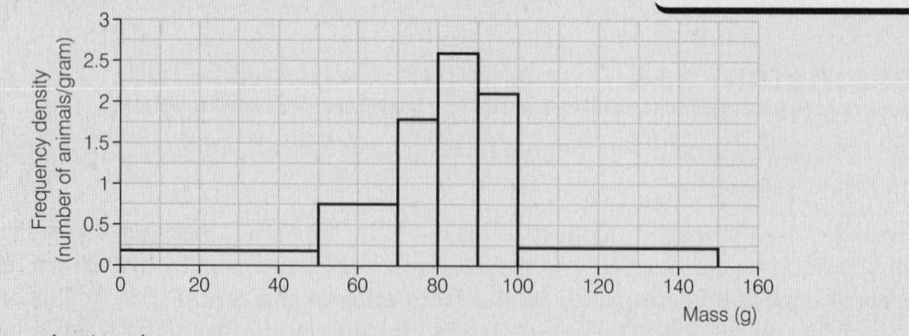

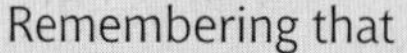

Remembering that

areas under a histogram represent frequencies

an estimate of the number of animals in the sample with a mass between 53 and 78 grams can be calculated by finding the area under the histogram between 53 and 78:

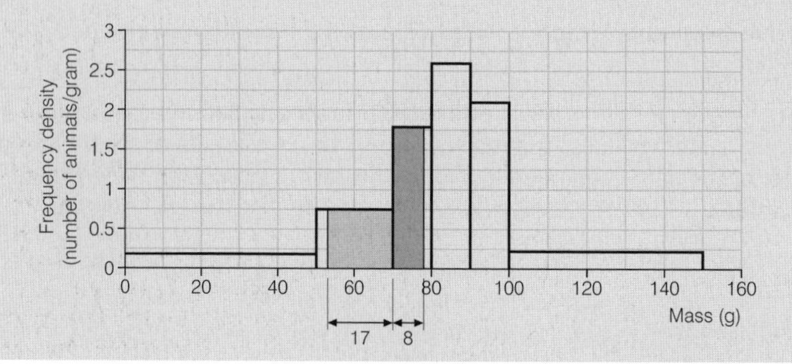

EXAMPLE 8 (continued)

Estimate of number of animals $= \text{area under graph}$
$= 17 \times 0.75 + 8 \times 1.8 \approx 27.$

You know from your frequency density calculations that the first shaded rectangle has height 0.75 and the second has height 1.8.

b) The mean mass is estimated using the midmarks of each group as the "x" values:

Mass	Frequency f	Midmark x	xf
0–50	9	25	225
50–70	15	60	900
70–80	18	75	1350
80–90	26	85	2210
90–100	21	95	1995
100–150	11	125	1375
	100		8055

So

$$\text{Mean} = \frac{8055}{100} = 80.55 \text{ grams}$$

You would normally do this on your calculator but remember that your working should include the main details:

Midmarks 25, 60, 75, 85, 95, 125

$$\bar{x} = \frac{\sum xf}{N} = \frac{8055}{100} = 80.55\,\text{g}$$

c) The **median** is usually found from a cumulative frequency diagram.

Mass	Frequency	Cumulative frequency	
0–50	9	9 animals with mass less than 50 g	
50–70	15	24	70 g
70–80	18	42	80 g
80–90	26	68	90 g
90–100	21	89	100 g
100–150	11	100	150 g

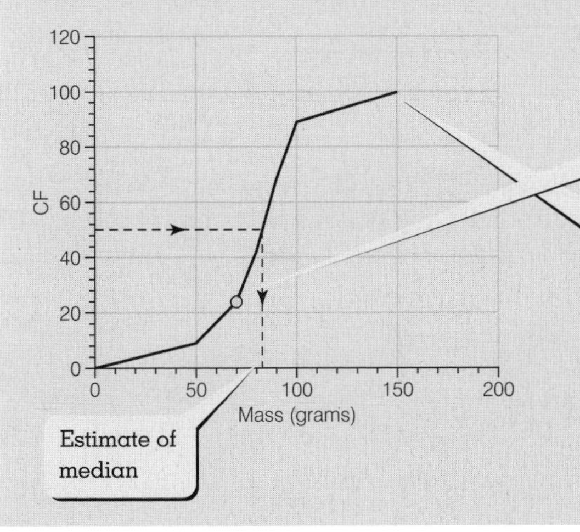

Remember to use the **upper boundary** of a group when plotting a CF graph. There were 24 animals with a mass less than 70 grams so you can plot a CF of 24 against a mass of 70 grams.

If there is a large amount of data and it has been split into a large number of groups it may be possible to join the points of the cumulative frequency graph with a smooth curve, but when there is only a small amount of data or just a few groups it is best to join the points of the cumulative frequency graph with straight line segments.

EXAMPLE 8 (continued)

For grouped data with n observations it is usual just to use the $\frac{1}{2}n$th observation to estimate the median.

Median = mass of 50th animal
$\approx$ 83 grams.

The median tells you that half of the animals weigh more than 83 grams and half of the animals weigh less than 83 grams. It thus gives you an idea of the mass of a typical animal.

The cumulative frequency graph can also give information about the number of animals with masses more than or less than a certain amount. For example, from the graph you can see that a mass of 120 grams corresponds to a cumulative frequency of approximately 93. This means that approximately 93 animals had a mass **less than** 120 grams and approximately 7 animals had a mass greater than 120 grams.

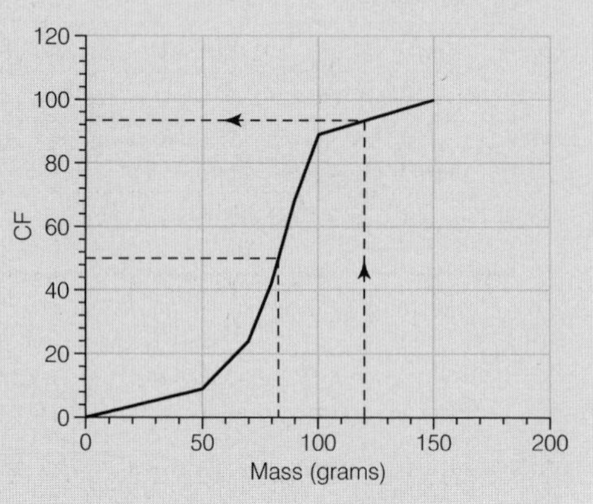

Care needs to be taken in the interpretation of histograms to ensure that the units of the frequency density are taken into account when calculating the frequencies of the groups.

EXAMPLE 9

The diagram is a histogram showing the values of the laptop computers owned by a university. How many laptop computers does the university own?

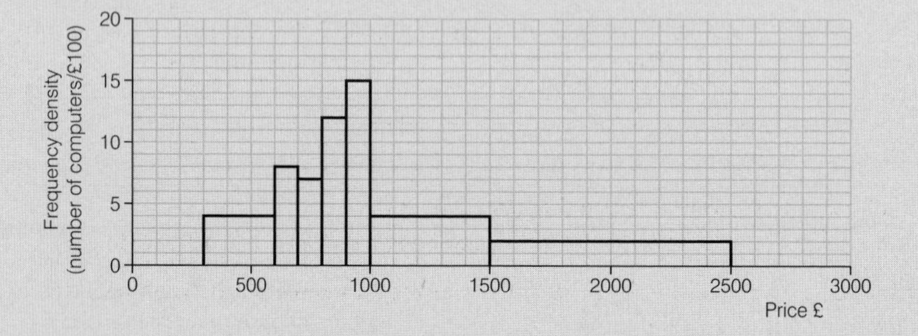

SOLUTION

The important factor in this example is the fact that the frequency density is given in terms of the number of computers/£100. This means that the first group (£300–£600) should be regarded as being 3 units wide, the second group (£600–£700) should be regarded as being 1 unit wide, etc.

Total number of computers = Area under the graph

$= 3 \times 4 + 1 \times 8 + 1 \times 7 + 1 \times 12 + 1 \times 15 + 5 \times 4 + 10 \times 2$

$= 94$

EXERCISE 2

1 The table below shows the speeds of cars passing a certain point on a motorway.

Speed (km/hr)	50–80	80–90	90–100	100–110	110–120	120–140
Frequency	12	9	15	37	32	24

a) Draw a histogram to illustrate the data and calculate an estimate of the number of cars travelling at between 93 and 117 km/hr.
b) Calculate an estimate of the mean speed of the cars passing this point.

2 The weight of portions of chips served at a school canteen were measured and gave the following data:

Weight (g)	0–50	50–100	100–150	150–200	200–300
Frequency	27	72	164	172	47

a) Draw a cumulative frequency graph on graph paper.
b) Use the graph you have drawn to find:

i) the median weight of the servings;
ii) the number of servings whose weight was more than 120 grams.

3 The students at a college did a survey into their travelling times to college and obtained the following results:

Travelling time (mins)	Number of students
0–10	3
10–20	7
20–30	12
30–40	25
40–50	29
50–60	18
60–70	10
70–80	6

a) Draw a suitable diagram to illustrate this data.
b) Calculate an estimate of the mean travelling time of these students.

4 The length of time taken by golfers to complete a round of 18 holes was recorded and summarised in the table below.

Time (mins)	150–	175–	200–	225–	250–275
Frequency	12	42	37	19	10

Draw a cumulative frequency graph to illustrate the data and hence obtain an estimate for the median time taken by the golfers to complete a round of golf.

5 The prices of 500 houses offered for sale on the south side of Manchester during a certain week of 2001 are summarised in the following table:

Price (£)	Frequency
0–25 000	0
25 000–75 000	62
75 000–100 000	130
100 000–125 000	90
125 000–150 000	107
150 000–200 000	70
200 000–300 000	21
300 000–400 000	20

a) Represent this information in a histogram.
b) Find the percentage of houses in the survey whose price lay between £80 000 and £120 000.
c) Calculate the mean price of the houses in the survey.

6 The diagram shows a histogram of the annual salaries of the employees at a certain factory.
a) How many employees are there at the factory?
b) What percentage of the employees earn between £11 500 and £12 500?

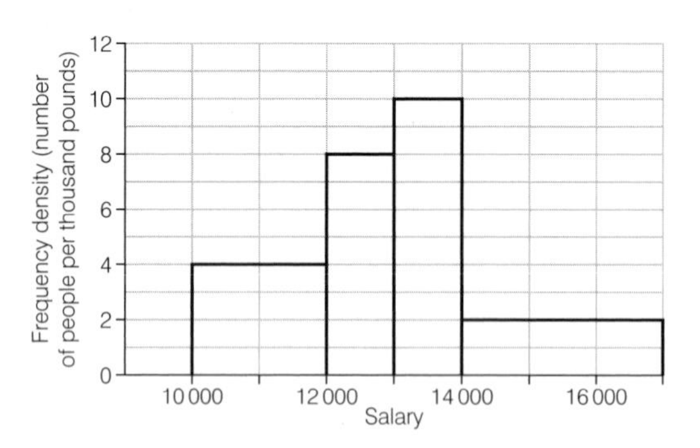

7 The diagram is a histogram showing the price of the used cars offered for sale at a large used car warehouse.
a) How many cars does the warehouse have for sale?
b) What proportion of the cars cost less than £3500?
c) Calculate an estimate of the mean price of the used cars at the warehouse.

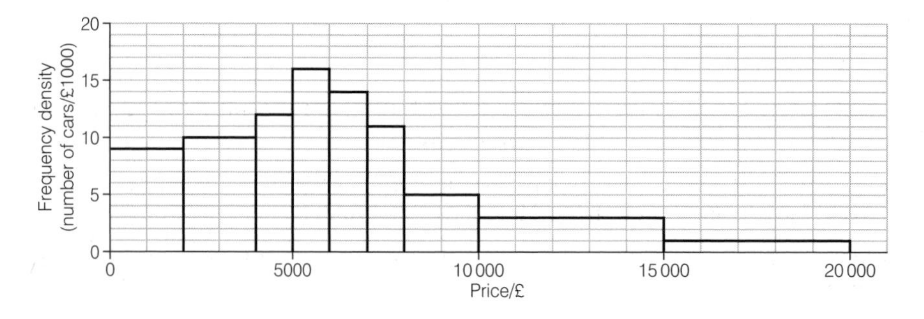

Interpreting Group Limits in Grouped Frequency Tables

Great care often needs to be taken when interpreting the information given in a grouped frequency table.

In the previous section, the data was split into non-overlapping but touching groups of the form

0–20 20–30 30–40 ...

where

0–20 takes all the data values that satisfy $0 \leqslant x < 20$
20–30 takes all the data values that satisfy $20 \leqslant x < 30$ etc.

Often this will not be the case, and you will be given groups which initially appear not to touch.

Consider for example the tasks of drawing a histogram, calculating the mean and drawing a cumulative frequency diagram to find the median for the masses, to the nearest kilogram, of 50 rugby players, summarised in the grouped frequency table below.

Mass (kg)	Frequency
70–79	2
80–89	4
90–94	1
95–99	10
100–104	9
105–109	8
110–114	4
115–124	9
125–134	3

The table shows the **group limits** and frequencies for each group. Notice that there is no ambiguity at all about the group limits.

The first important task is to interpret exactly the meaning of the group limits.

Since the masses are given to the nearest kilogram, a rugby player appearing in the 90–94 group has an **exact mass** that satisfies

$89.5 \leqslant \text{exact mass} < 94.5$

The numbers 89.5 and 94.5 are the **true group boundaries** of this group: 89.5 is the lower group boundary and 94.5 is the upper group boundary.

The **group width** is the difference between the two group boundaries. Thus the group width of the 90–94 group is 94.5 – 89.5 = 5 kg.

The **midmark** of the group is the average of the group boundaries of the group. Thus the midmark of the 90–94 group is

$$\frac{89.5 + 94.5}{2} = 92$$

These ideas are illustrated in the following number line diagram.

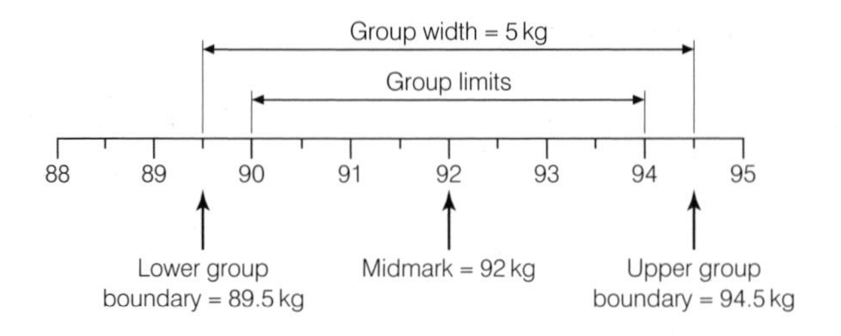

In all analyses of data it is important that the true group boundaries are used rather than the group limits.

Thus the histogram will have bars going from 69.5 to 79.5, from 79.5 to 89.5, etc.

Mass (kg)	Mass (kg) true group boundaries	Frequency	Frequency density
70–79	**69.5–79.5**	2	0.2
80–89	**79.5–89.5**	4	0.4
90–94	**89.5–94.5**	1	0.2
95–99	**94.5–99.5**	10	2.0
100–104	**99.5–104.5**	9	1.8
105–109	**104.5–109.5**	8	1.6
110–114	**109.5–114.5**	4	0.8
115–124	**114.5–124.5**	9	0.9
125–134	**124.5–134.5**	3	0.3

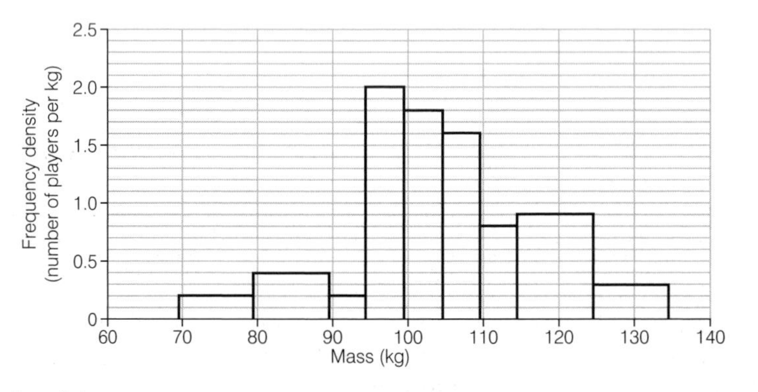

The midmarks of the groups are

4.5, 84.5, 92, 97, 102, 107, 112, 119.5, 129.5

so

$$\bar{x} = \frac{\sum xf}{N} = \frac{5235}{50} = 104.7 \text{ kg}$$

Check this on your calculator.

The cumulative frequency table for the data is

Mass (kg)	Mass (kg) true group boundaries	Frequency	Cumulative frequency
70–79	**69.5–79.5**	2	2 with mass less than 79.5 kg
80–89	**79.5–89.5**	4	6 89.5 kg
90–94	**89.5–94.5**	1	7 94.5 kg
95–99	**94.5–99.5**	10	17 99.5 kg
100–104	**99.5–104.5**	9	26 104.5 kg
105–109	**104.5–109.5**	8	34 109.5 kg
110–114	**109.5–114.5**	4	38 114.5 kg
115–124	**114.5–124.5**	9	47 124.5 kg
125–134	**124.5–134.5**	3	50 134.5 kg

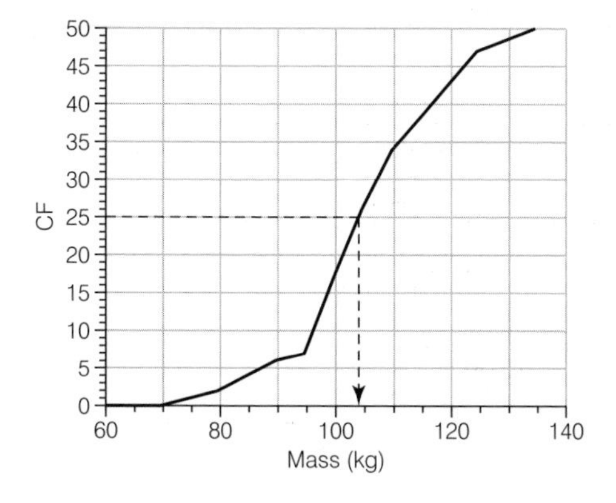

Median = 25th piece of data ≈ 104 kg.

EXAMPLE 10

Consider the following set of data which shows the ages, in completed years, of the audience at a theatre:

Age (years)	5–19	20–29	30–39	40–49	50–69	70–99
Frequency	156	142	167	198	142	67

a) Draw a histogram to illustrate the ages of the audience at the theatre. How many of the audience were at least 16 years old but less than 25 years old?

b) Calculate an estimate for the mean age of the audience.

SOLUTION

a) We normally measure age in **completed years**. Thus the girl who is 7 years and 323 days old will say that she is 7 years old.

The 5–19 group will therefore contain all the people whose exact age satisfies

$$5 \leqslant \text{exact age} < 20$$

and the midpoint of this group is 12.5 years.

EXAMPLE 10 (continued)

Similarly, the 20–29 group will contain all the people whose exact age satisfies

$$20 \leqslant \text{exact age} < 30$$

and the midpoint of this group is 25 years.

Use the true group boundaries in **all** calculations.
Group width = 20 – 5.
Frequency density = $\frac{156}{15}$ = 10.4.

Age (years)	Age (years) true group boundaries	Frequency f	Frequency density
5–19	**5–20**	156	10.4
20–29	**20–30**	142	14.2
30–39	**30–40**	167	16.7
40–49	**40–50**	198	19.8
50–69	**50–70**	142	7.1
70–99	**70–100**	67	2.2333 ...

Notice there are no gaps between the true upper group boundary of one group and the true lower group boundary of the next group.

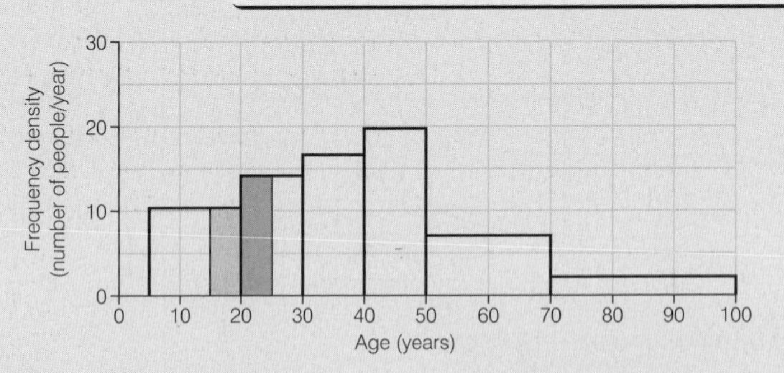

Estimate of number of people between 16 and 25 = Area of graph between 16 and 25

$= 4 \times 10.4 + 5 \times 14.2$

$= 112.6$

So, approximately 113 members of the audience were between 16 and 25 years old.

b) The midmarks of the groups are 12.5, 25, 35, 45, 60 and 85, so

$$\bar{x} = \frac{\Sigma xf}{N} = \frac{34470}{872} = 39.53 \text{ years} \qquad (2 \text{ d.p.})$$

EXAMPLE 11

The distribution of exam marks obtained by a group of students is given in the following table:

Mark	10–39	40–49	50–59	60–79	80–99
Frequency	7	21	39	21	12

Draw a cumulative frequency diagram for the data and hence find the median of the marks.

EXAMPLE 11 (continued)

SOLUTION

An exam mark is often a whole number, so the 40–49 group will contain all the pupils whose marks are

40, 41, 42, 43, 44, 45, 46, 47, 48 or 49

but it is possible that the marks may have been rounded to the nearest whole number, so it is usual to think of the 40–49 group as containing all pupils whose exact mark satisfied

$39.5 \leqslant \text{exact mark} < 49.5$

Mark	Mark true group boundaries	Frequency f	Cumulative frequency
10–39	**9.5–39.5**	7	7 students with a mark less than 39.5
40–49	**39.5–49.5**	21	28 49.5
50–59	**49.5–59.5**	39	67 59.5
60–79	**59.5–79.5**	21	88 79.5
80–99	**79.5–99.5**	12	100 99.5

Notice there are no gaps between the true upper group boundary of one group and the true lower group boundary of the next group.

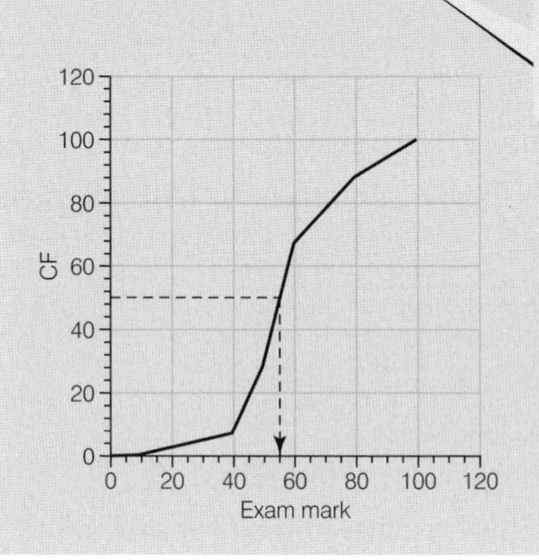

The median is the mark obtained by the 50th student and it can be seen from the graph that the median can be estimated as 55 marks.

Estimating the Median by Linear Interpolation

It is possible to estimate the median of grouped data quickly using some ratio theory.

Consider again the data of Example 8 giving the masses of 100 small animals.

Mass	Frequency	Cumulative frequency
0–50	9	9 animals with mass less than 50 g
50–70	15	24 animals with mass less than 70 g
70–80	18	42 animals with mass less than 80 g
80–90	26	68 animals with mass less than 90 g
90–100	21	89 animals with mass less than 100 g
100–150	11	100 animals with mass less than 150 g

The median is the mass of the 50th animal. Looking at the cumulative frequencies we can see that this lies between 80 and 90 grams.

50 is $\frac{8}{26}$ of the way from 42 to 68: the median will therefore be approximately $\frac{8}{26}$ of the way from 80 to 90 and we can therefore write

$$\text{Median} \approx 80 + \tfrac{8}{26} \times 10 \approx 83.1 \text{ g}$$

Alternatively, the estimate of the median can be found by sketching the part of the CF graph for the 80–90 group where the median is known to lie.

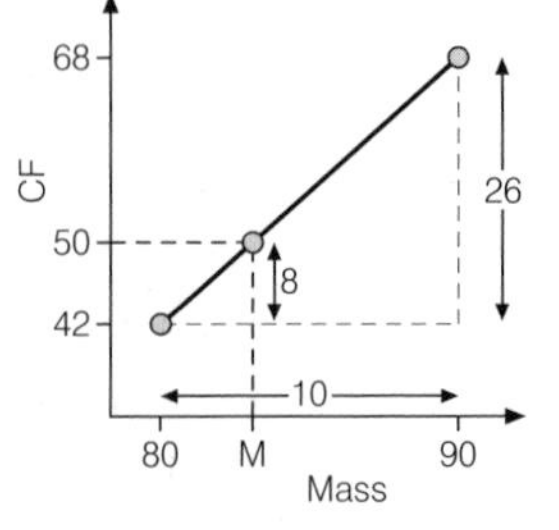

Gradient PQ = Gradient PR

$\Rightarrow \quad \frac{26}{10} = \frac{8}{M - 80}$

$\Rightarrow \quad M - 80 = 8 \times \frac{10}{26}$

$\Rightarrow \quad M = 80 + \frac{8}{26} \times 10 \approx 83.1 \text{ g}$

Notice that this method assumes the points on the cumulative frequency graph have been joined with straight lines rather than with a curve.

EXERCISE 3

1 The table below shows the annual incomes, in thousands of pounds, of the inhabitants of a village who are in full-time employment.

Income	0–	10–	15–	20–	30–50
Frequency	82	156	182	112	63

a) Draw a histogram to illustrate the distribution of incomes.
b) Calculate the mean annual income of the inhabitants of this village.

The diagram below shows the histogram of annual incomes of the inhabitants of a second village who are in full-time employment. The estimate of the mean income of these people has been calculated as £29 900, correct to three significant figures.

Use the two histograms and the estimates of the mean annual incomes to write a brief paragraph comparing the distributions of incomes in the two villages.

2 The table below shows the length, correct to the nearest minute, of films shown at a cinema during the course of a year.

Time (mins)	60–99	100–109	110–119	120–149	150–179
Frequency	7	9	13	8	4

a) Write down the true group boundaries and the midmark of the 60–99 group.
b) Calculate an estimate of the mean length of these films.
c) Use linear interpolation to calculate an estimate of the median length of the films.

3 The table below shows the ages of the audience at a concert.

Age (years)	10–19	20–24	25–29	30–44	45–74
Frequency	82	65	95	83	27

a) Write down the true group boundaries of the 10–19 group.
b) Draw a cumulative frequency graph to illustrate the data and hence estimate:

i) the median age of this audience;
ii) the percentage of the audience aged less than 35.

4 The table below shows the heights, to the nearest centimetre, of saplings after six months of growth.

Height (cm)	5–14	15–19	20–24	25–29	30–39	40–59
Frequency	3	26	16	9	8	6

a) Write down the true group boundaries and the midmark of the 5–14 group.
b) Draw a histogram to illustrate the data.
c) Use linear interpolation to calculate an estimate for the median height of the saplings.
d) Calculate an estimate for the mean height of the saplings.
e) Explain the reason for different values you have obtained for the "average" height of a sapling.

5 A survey was conducted to investigate the length of time adults spent listening to a radio each week. The data collected is illustrated in a cumulative frequency graph.

a) Find an estimate for the median time spent listening to the radio by these adults.
b) Obtain the grouped frequency table for the data and hence calculate an estimate of the mean time spent listening to the radio by these adults.

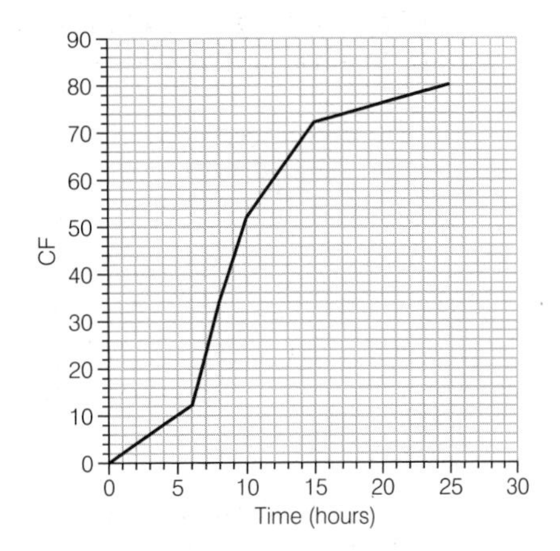

6 The data below shows the gaps, in seconds, between vehicles travelling along a quiet country lane.

90, 74, 49, 40, 112
137, 37, 83, 60, 100
15, 104, 12, 98, 18
20, 13, 53, 46, 21
103, 74, 82, 96, 108
12, 15, 17, 7, 233
85, 130, 7, 16, 7
76, 35, 13, 34, 79
40, 14, 23, 6, 11
19, 63, 5, 87, 274

a) Calculate the mean of this data.

b) Summarise the data as a grouped frequency table using groups of 0–29, 30–59, 60–89, 90–119, 120–179, 180–239, 240–299 and draw a histogram to illustrate this grouped frequency distribution.

c) Calculate the mean of this grouped frequency distribution. Explain why your answer is different to the answer obtained in (a).

Having studied this chapter you should know how to

- represent qualitative data using a bar chart or a pie chart
- represent discrete ungrouped data using a frequency bar chart, frequency polygon or vertical line diagram
- represent grouped data using a histogram, which is a bar chart where the heights of the bars show the frequency density of each group and can be calculated from using

$$\text{Frequency density} = \frac{\text{Frequency}}{\text{Group width}}$$

- estimate frequencies from a histogram using the fact that the areas of the bars of a histogram give the frequencies of the groups
- calculate the mean and median of ungrouped data
- estimate the mean of grouped data using the midmarks of each group
- estimate the median of grouped data using either a cumulative frequency graph or linear interpolation
- use sigma notation to express the mean, $\bar{x}$, algebraically. In the case when each of the observations $x_1, x_2, x_3, \ldots, x_n$ occurs with frequency 1

$$\bar{x} = \frac{\Sigma x}{n}$$

In the case when the observations $x_1, x_2, x_3, \ldots, x_n$ occur with frequencies $f_1, f_2, f_3, \ldots, f_n$, respectively

$$\bar{x} = \frac{\Sigma xf}{N} \text{ where } N = \Sigma f$$

REVISION EXERCISE

1 Keiko has conducted an investigation to determine whether the time that she uses the internet has a marked effect on the time that it takes for new pages to be loaded.

She accessed the same 60 pages on a Thursday evening and on a Sunday morning. The results for Thursday evening are summarised in the table below.

Time (s)	0–	10–	20–	30–	40–	60–	80–	100–200
Frequency	14	11	9	7	6	5	5	3

Draw a histogram to illustrate the data and calculate an estimate of the mean loading time for an internet page on Thursday evening.

The diagram below is a histogram of her results for Sunday morning.

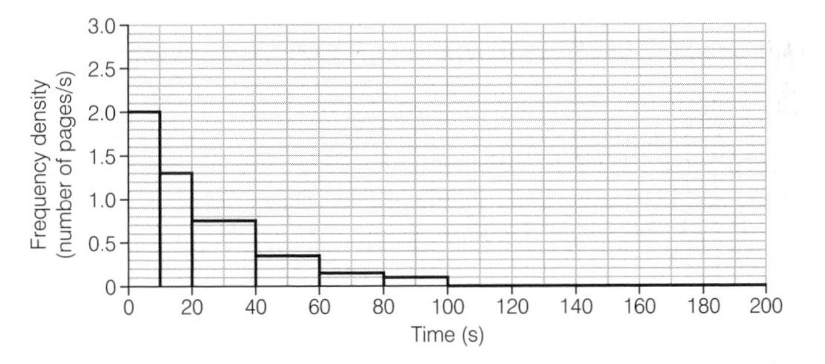

Obtain the grouped frequency table for Sunday morning and calculate an estimate of the mean of this distribution.

What conclusions can be drawn from Keiko's investigation?

2 The number of hours spent watching television during a week by a group of 11-year-olds was recorded *correct to the nearest hour* and yielded the following results:

Time (hr)	0–4	5–7	8–10	11–13	14–20
Frequency	12	31	57	65	41

Draw a cumulative frequency graph for the data and hence estimate:

a) the median time spent watching TV by these children;

b) the percentage of children who watch more than eight and a half hours TV a week.

3 The table below shows the ages, *in completed years*, of the pupils at a school on the 30th July 2003.

Age (yr)	7–11	12–13	14	15	16	17–18
Frequency	83	127	87	79	84	159

a) Draw a histogram to illustrate the data.

b) Calculate the mean age of the pupils in the school.

4 A set of data is summarised by

$$\Sigma xf = 4530 \qquad \Sigma f = 60$$

Calculate the mean of the data.

5 The table below shows the times taken by the participants in a 10-km fun run.

Time (mins)	30–	50–	60–	70–	90–	120–150
Frequency	19	83	112	82	65	43

a) Draw a cumulative frequency graph to illustrate the data.
b) Use the graph to estimate the median time of the participants.
c) Estimate the number of runners whose average speed for the fun run was greater than 8 km/hr.

6 The information below gives the number of goals scored in the Premiership and the Coca-Cola Championship, League 1 and League 2 on one Saturday.

1 3 4 6 2   2 1 3 3 1   5 3 6 5 0   3 1 2 2 4   2 2 4 4 7
4 1 1 4 3   2 4 2 0 2   2 6 2 4 0   9

a) Illustrate the information in a suitable diagram.
b) Calculate the mean and median number of goals per match.

7 The table below shows the marks obtained in an IQ test by the pupils at a school.

IQ mark	60–89	90–99	100–109	110–119	120–129	130–159
Frequency	5	12	72	65	31	12

a) Draw a histogram to illustrate this set of data.
b) Calculate an estimate of the mean IQ score of pupils at the school.

8 The diagram is a histogram showing the ages of the employees of a large department store.

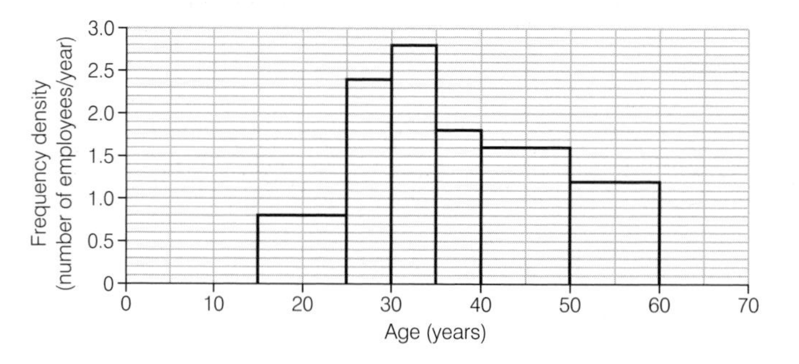

a) How many employees does the department store have?
b) Estimate the percentage of employees that are over 42 years old.

9 A sample of 500 electric light bulbs was tested to see how many continuous hours they would work for.

Time (hr)	500	700	900	1100	1500
Number of bulbs that had failed by this time	23	178	321	441	500

a) By drawing a suitable graph, obtain the median lifetime of these bulbs.
b) Copy and complete the following table giving the frequency distribution of the lifetimes of these bulbs:

Time (hr)	0–500	500–700			
Frequency	23	155			

and hence:

i) draw a histogram for the data;
ii) calculate an estimate of the mean lifetime of the bulbs.

10 A survey was conducted to discover how much a family of four spends on their main annual holiday. The results are illustrated in a cumulative frequency diagram.

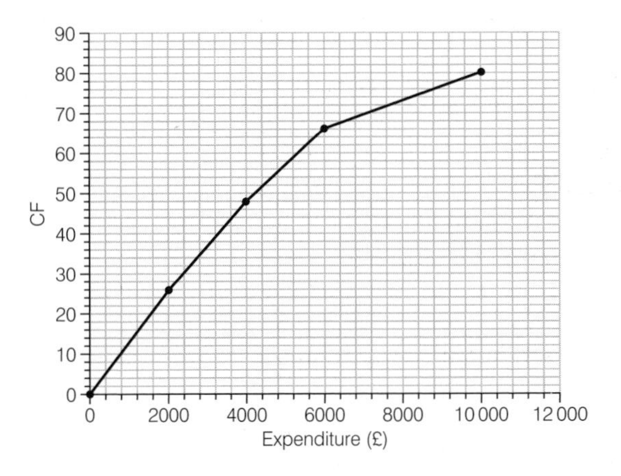

a) Estimate the median of this data.
b) Summarise the data in a grouped frequency table.
c) Estimate the mean expenditure of a family of four on their main annual holiday.

2 Presentation of Statistical Data: Measures of Spread

The purpose of this chapter is to enable you to

- understand the need for measures of spread
- find the range of a distribution
- find the interquartile range of grouped and ungrouped data
- use stem and leaf diagrams
- use box and whisker diagrams as a method of comparing distributions
- use the variance and standard deviation as measures of spread

The Need for Further Measures

The previous chapter demonstrated how to calculate the averages or measures of central tendency of a frequency distribution. However, these statistics give only a partial summary of a distribution.

Consider, for example, a class of 25 pupils who had three pieces of homework to do one evening. They recorded the time that each subject took and summarised the results in the table below.

Time (mins)	Frequency for French	Frequency for Maths	Frequency for History
$10 \leqslant T < 20$	0	5	3
$20 \leqslant T < 30$	6	5	9
$30 \leqslant T < 40$	13	5	4
$40 \leqslant T < 50$	6	5	3
$50 \leqslant T < 60$	0	5	6

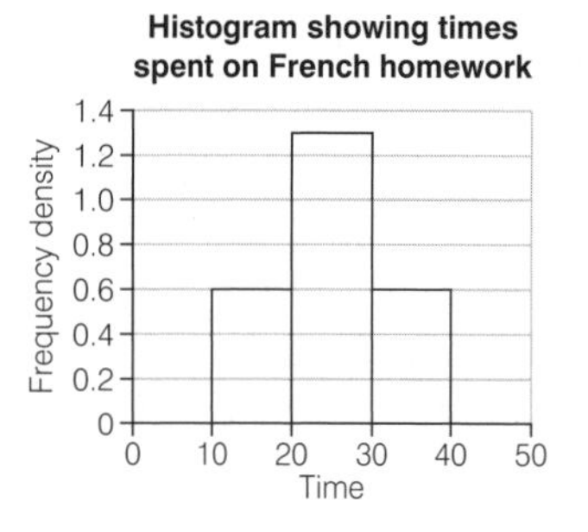

For the French times

$$\bar{x} = \frac{\Sigma xf}{N}$$

$$= \tfrac{625}{25} = 25\,\text{min}$$

For the Maths times

$$\bar{x} = \frac{\Sigma xf}{N}$$

$$= \tfrac{625}{25} = 25\,\text{min}$$

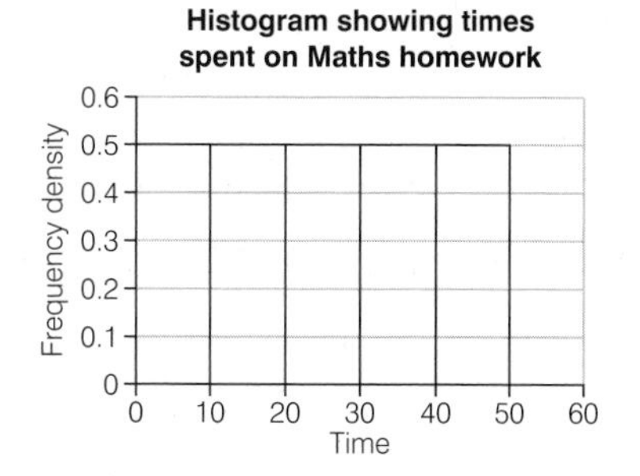

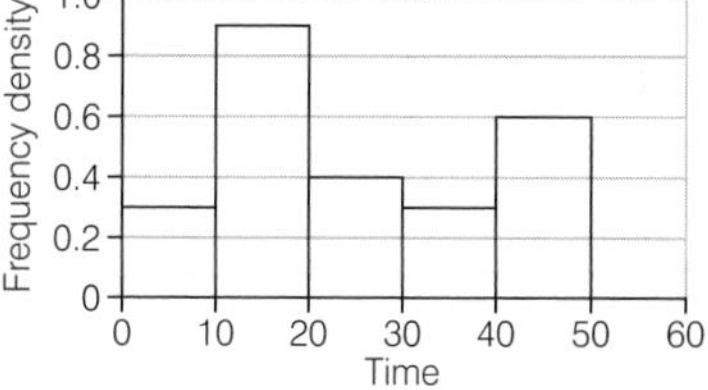

For the History times

$$\bar{x} = \frac{\Sigma xf}{N}$$

$$= \tfrac{625}{25} = 25\,\text{min}$$

Although the mean time spent on each subject was exactly the same, you can see from the frequency tables and the histograms that the distributions are very different.

This suggests that the mean does not adequately summarise a frequency distribution and that it is necessary to introduce other measures or statistics to further summarise the distribution. As well as calculating averages, we need to calculate measures that give information about the **spread** of the distribution.

Range and Interquartile Range

The simplest measure of the spread of a set of data is the **range**. This is the difference between the largest and smallest observations:

range = largest observed value – smallest observed value

The range is not a particularly satisfactory measure of spread. It is difficult to evaluate precisely once data have been grouped. For example, the range of times spent on history homework could be as much as 60 – 10 = 50 minutes or as little as 50 – 20 = 30 minutes or anything in between, depending on the precise observations.

The range is also totally determined by the extreme values of the data, which may not be typical of the data as a whole. For example, if the weekly earnings of the 10 workers in a small factory were £200, £200, £200, £200, £200, £200, £250, £250, £250, £950 then the range would be £750, which suggests a large spread of wages even though most of the wages are in a small £50 interval.

This problem can be overcome by calculating a measure of spread known as the **interquartile range**. This is the range of the middle 50% of the distribution and its value is therefore independent of any exceptional extreme values.

The interquartile range is calculated by:

- first finding the **upper quartile** (or 75th percentile) of the distribution, which is the value taken by the observation 3/4 of the way along the data when the data are listed in order;
- then finding the **lower quartile** (or 25th percentile) of the distribution, which is the value taken by the observation 1/4 of the way along the data when the data are listed in order;
- and finally calculating the difference between the two quartiles.

interquartile range = upper quartile – lower quartile.

Estimating the Interquartile Range for Grouped Data

EXAMPLE 1

Find the interquartile range of the masses of the animals given in the table below.

Mass (grams)	0–	50–	70–	80–	90–	100–150
Frequency	9	15	18	26	21	11

SOLUTION

A cumulative frequency diagram can be used to find the **interquartile range**:

Mass	Frequency	Cumulative frequency
0–50	9	9 animals with mass less than 50 g
50–70	15	24 ... 70 g
70–80	18	42 80 g
80–90	26	68 90 g
90–100	21	89 100 g
100–150	11	100 150 g

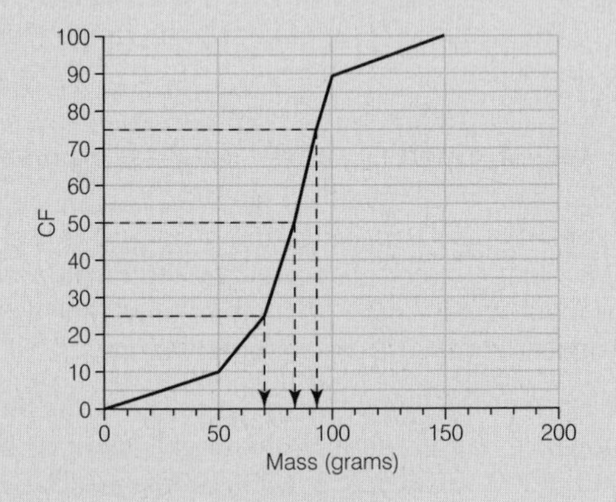

Lower quartile = mass of 25th animal ≈ 70 grams
Median = mass of 50th animal ≈ 83 grams
Upper quartile = mass of 75th animal ≈ 93 grams

Interquartile range = upper quartile – lower quartile ≈ 93 – 70 = 23 grams

EXAMPLE 1 (continued)

The interquartile range can also be calculated using linear interpolation:

Lower quartile $\approx 70 + \frac{1}{18} \times 10 \approx 70.6$ g

Upper quartile $\approx 90 + \frac{7}{21} \times 10 \approx 93.3$ g

$\Rightarrow$ Interquartile range $\approx 93.3 - 70.6 = 22.7$ g

25 is $\frac{1}{18}$ of the way from 24 to 42 so the lower quartile is $\frac{1}{18}$ of the way from 70 to 80, giving lower quartile $= 70 + \frac{1}{18} \times 10$.

EXAMPLE 2

The table below shows the prices of a sample of 120 two-bedroomed houses for sale in each of the cities of Northton and Southville.

Price (thousands of pounds)	50–	75–	100–	125–	150–	200–250
Northton frequencies	23	47	28	15	7	0
Southville frequencies	7	22	38	29	18	6

Illustrate the two distributions on a single cumulative frequency diagram and determine the medians and interquartile ranges of each distribution.

SOLUTION

The diagram shows the cumulative frequency graphs for the two cities.

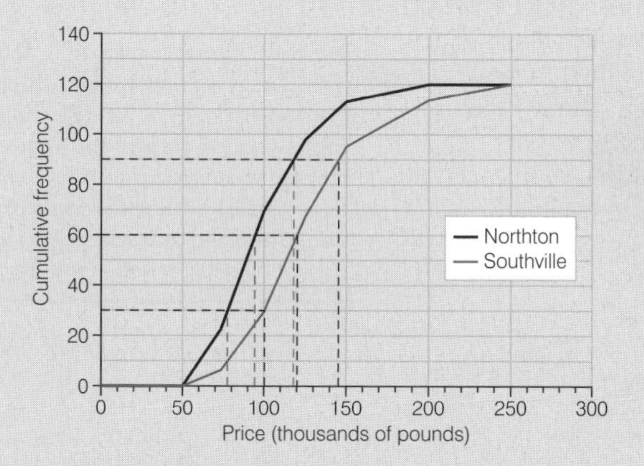

Since there are 120 houses in each city:

Lower quartile = 30th house
The lower quartile for Northton is approximately £79 000
The lower quartile for Southville is approximately £101 000

Median = 60th house
The median for Northton is approximately £95 000
The median for Southville is approximately £120 000

Upper quartile = 90th house
The upper quartile for Northton is approximately £118 000
The upper quartile for Southville is approximately £145 000

Interquartile range for Northton $\approx$ £118 000 – £79 000 = £39 000
Interquartile range for Southville $\approx$ £145 000 – £101 000 = £44 000

Box Plots (or Box and Whisker Plots)

In the last example, both the initial frequency distributions and the cumulative frequency diagram make it clear that the houses in Southville are generally more expensive than the houses in Northton. A **box plot** provides a quick and easy way of comparing different distributions in more detail.

Along a number line, a box is formed by the upper and lower quartiles. The median divides the box into two portions. The whiskers extend from the box to the maximum and minimum values of the data.

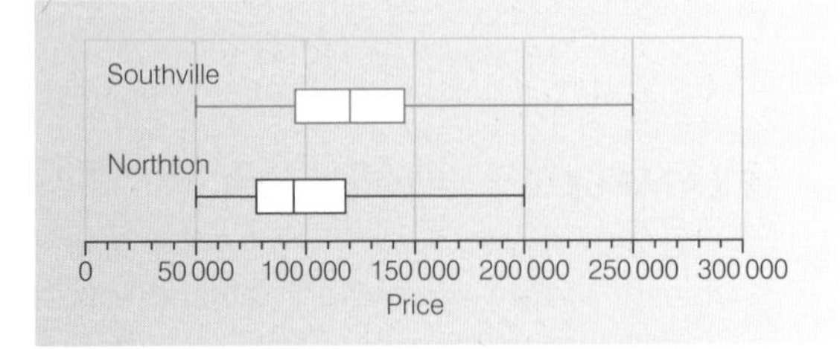

From the box plot it is possible not only to make the general comment that the houses in Northton are cheaper than the houses in Southville, but more precise comments can be made. For example:

- the average (median) house in Northton is approximately £25 000 cheaper than the average house in Southville;
- over 50% of the houses for sale in Northton but less than 25% of the houses for sale in Southville cost less than £100 000;
- the spread (range) of house prices in Southville is much greater than the range of the prices in Northton, but the central 50% of houses in each town have approximately equal spreads (interquartile range).

Although box plots give an instant picture of the distribution with information about the three quartiles and the extremes of the distribution, enabling useful comparisons between distributions, cumulative frequency diagrams do contain more information. For example, the 60th percentile of a distribution is easily estimated from a cumulative frequency diagram but cannot be satisfactorily estimated from a box plot.

EXERCISE 1

1 The table below shows the annual incomes, in thousands of pounds, of the inhabitants of a town who are in full-time employment.

Income	0–15	15–20	20–25	25–40	40–70
Frequency	82	156	182	112	63

a) By drawing a cumulative frequency diagram, or otherwise, obtain estimates of:

i) the median income of these people;
ii) the interquartile range of the incomes.

b) Draw a box and whisker diagram to illustrate the data.

2 A group of slimmers followed a new diet for a period of three months and their weight losses, to the nearest pound, are shown in the table below.

Weight loss (lb)	0–4	5–9	10–14	15–19	20–24
Frequency	20	32	41	20	7

a) Draw a cumulative frequency diagram for the data and hence estimate the median and interquartile range of the weight losses.
b) Draw a box plot to illustrate the data.

3 The table below shows the ages, in completed years, of the inhabitants of a South Wales town.

Age	0–14	15–29	30–44	45–59	60–74	75–99
Frequency	812	1189	1215	1077	1567	740

a) By drawing a cumulative frequency graph, or otherwise, determine the median and interquartile range of these ages.
b) Illustrate the data with a box plot (otherwise known as a box and whisker plot).

The diagram below shows a box plot for the ages of the inhabitants of a town in the West Midlands of England.

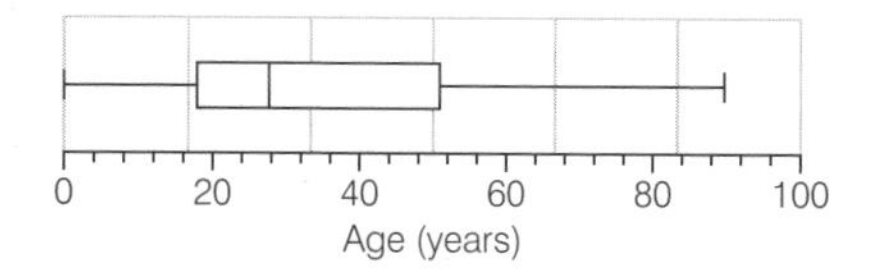

Write a short paragraph comparing the age distributions of the inhabitants of the two towns.

4 In an examination, the marks obtained by the candidates were such that:

the lowest score obtained was 18 marks;
20% of the candidates scored at most 30 marks;
40% of the candidates scored at most 46 marks;
60% of the candidates scored at most 58 marks;
80% of the candidates scored at most 78 marks;
the highest mark obtained was 95 marks.

a) Calculate, making your method clear, estimates for:

i) the median mark in the examination;
ii) the interquartile range of the marks in the examination.

b) Draw a box plot for the data.

Interquartile Range for Ungrouped Data

The median of n ungrouped observations is the $\frac{1}{2}(n+1)$th observation when the data have been ordered, and if n is even this will involve finding the average of two observations. Similarly, the lower quartile of n ungrouped observations is the $\frac{1}{4}(n+1)$th observation, and the upper quartile is the $\frac{3}{4}(n+1)$th observation.

EXAMPLE 3

Find the interquartile range of the 11 observations

25, 18, 16, 41, 36, 23, 52, 32, 28, 41, 49

SOLUTION

In order the observations are

16 18 23 25 28 32 36 41 41 49 52

$\frac{1}{4}(11+1)=3$, so

lower quartile = 3rd observation = 23

$\frac{3}{4}(11+1)=9$, so

upper quartile = 9th observation = 41

Interquartile range = upper quartile – lower quartile = 41 – 23 = 18

Usually, $\frac{1}{4}(n+1)$ and $\frac{3}{4}(n+1)$ will not be integers. The following example shows how the quartiles can be determined in this case.

EXAMPLE 4

Find the interquartile range of the 10 observations

25, 18, 16, 41, 36, 23, 52, 32, 28, 41

SOLUTION

In order the observations are

16 18 23 25 28 32 36 41 41 52

$\frac{1}{4}(10+1)=2.75$, so

lower quartile = 2.75th observation = 21.75

Take the second observation, which is 18, and the third, which is 23.
The "2.75th" observation is $\frac{3}{4}$ of the way from 18 to 23:

$$\frac{3}{4}\times(23-18)=3.75$$

The "2.75th" observation is $18+3.75=21.75$.

$\frac{3}{4}(10+1)=8.25$, so

upper quartile = 8.25th observation = 41

The 8th and 9th observations are both 41 so the "8.25th" observation is 41.

Interquartile range = upper quartile – lower quartile = 41 – 21.75 = 19.25

The interquartile range for ungrouped data that is already summarised in a frequency table can be calculated by adding a column of cumulative frequencies to the table.

EXAMPLE 5

The table shows the number of days absence that a group of students had during a term. Find the median and interquartile range.

Number of days absence	0	1	2	3	4	5	6	7
Frequency	1	4	4	5	6	7	2	1

EXAMPLE 5 (continued)

SOLUTION

Adding the cumulative frequencies to the table gives

Number of days absence	Frequency	Cumulative frequency
0	1	**1** student had 0 days absence
1	4	**5** students had 1 day or less absence
2	4	**9** students had 2 days or less absence
3	5	**14** students had 3 days or less absence
4	6	**20** students had 4 days or less absence
5	7	**27** students had 5 days or less absence
6	2	**29** students had 6 days or less absence
7	1	**30** students had 7 days or less absence

There are 30 observations altogether.

$\frac{1}{2}(30 + 1) = 15.5$, so the median is the average of the 15th and 16th observations when the data are taken in order. From the table we see that the 15th and 16th students both had 4 days absence, so the median is 4 days.

For the lower quartile, $\frac{1}{4}(30 + 1) = 7.75$, so the 7th and 8th observations need to be taken into account. The 7th and 8th students both had 2 days absence, so the lower quartile is 2 days.

For the upper quartile, $\frac{3}{4}(30 + 1) = 23.25$, so the 23rd and 24th observations need to be taken into account. The 23rd and 24th students both had 2 days absence, so the upper quartile is 5 days.

Interquartile range = upper quartile – lower quartile = 5 – 2 = 3 days.

Stem and Leaf Diagrams

For ungrouped data that have not been summarised as a frequency table, a **stem and leaf** diagram provides an alternative method of summarising the data. The frequency distribution and the median and interquartile range can rapidly be deduced from the stem and leaf diagram.

For a positive integer with two or more digits, the units digit of the number can be called the **leaf** of the number, whilst the other digits form the **stem**. For the number 382, 2 is the leaf and 38 is the stem.

To produce a stem and leaf diagram for a set of data:

- find the maximum and minimum observation values for the data, and use these values to write down all the possible stems for the data in the first column of a table;
- work through the data systematically, writing the leaf of each observation in the row belonging to the stem of the observation to produce the first-pass, or unsorted, stem and leaf diagram;
- produce the **final stem and leaf diagram** by sorting the leafs in each row into numerical order and putting a key or legend at the bottom of the diagram so that it can be easily interpreted.

This process is illustrated in the next example.

EXAMPLE 6

The numbers of goals scored by Premier League clubs during a football season were

74 85 63 68 61   52 48 43 47 51
48 45 41 41 58   42 41 42 29 21

Produce a stem and leaf diagram for the data and hence find the median and interquartile range of the data. Summarise the data in a box plot.

SOLUTION

The smallest number of goals scored was 21 and the largest number of goals scored was 85, so we need stems of 2, 3, 4, 5, 6, 7 and 8:

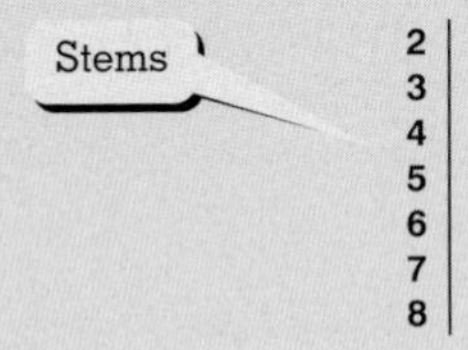

The first-pass, or unsorted, stem and leaf diagram is now obtained by working through the data systematically. The first team scored 74 goals, so you can put a 4 in the row which has a stem value of 7; the second team scored 85 goals, so you can put a 5 in the row which has a stem value of 8; the third team scored 63 goals, so you can put a 3 in the row which has a stem value of 6; the fourth team scored 68 goals, so you can put an 8 in the row which has a stem value of 68; etc.

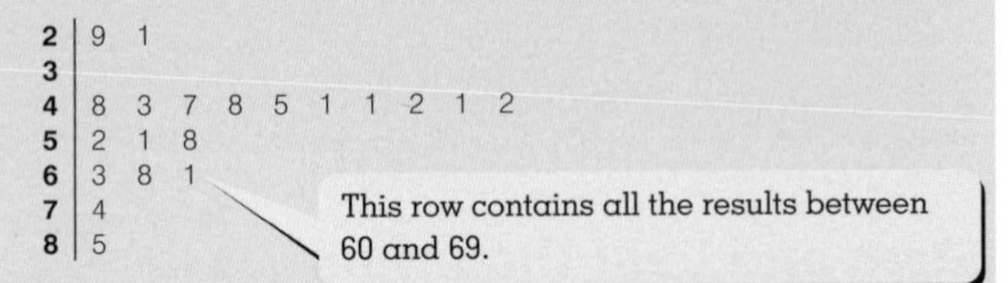

The final stem and leaf diagram completes the ordering of the data and includes a "key" so that the diagram can be readily understood.

Observe that the stem and leaf diagram gives a visual interpretation of the distribution. It resembles a histogram with groups of 20–, 30–, 40–, etc. drawn with horizontal rather than vertical bars. The stem and leaf diagram has the great advantage that the original data is immediately retrievable from the diagram.

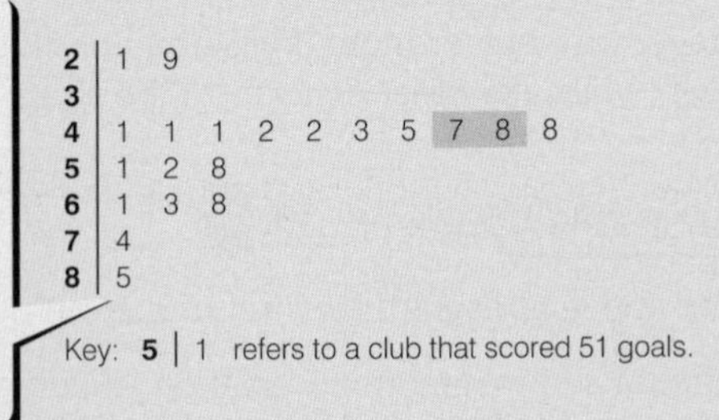

Median $= \frac{1}{2}(20 + 1)$th observation
$= 10.5$th observation = average of 10th and 11th observations
$= \frac{47 + 48}{2} = 47.5$ goals

Start counting from the top of the stem and leaf diagram: 21 is the first observation, 29 the second, 41 the third, ... The 10th and 11th observations are shaded in.

EXAMPLE 6 (continued)

For the lower and upper quartiles we use the $\frac{1}{4}(n+1)$th and $\frac{3}{4}(n+1)$th observations, respectively.

Lower quartile $= \frac{1}{4}(20+1)$th observation
$= 5.25$th observation
$= 41.25$

> Take the 5th observation, which is 41, and the 6th, which is 42.
> The "5.25th" observation is $\frac{1}{4}$ of the way from 41 to 42.
> The "5.25th" observation is 41.25.

Upper quartile $= \frac{3}{4}(20+1)$th observation
$= 15.75$th observation
$= 60.25$

> Take the 15th observation, which is 58, and the 16th, which is 61.
> The "15.75th" observation is $\frac{3}{4}$ of the way from 58 to 61.
> The "15.75th" observation is 60.25.

Interquartile range $= 60.25 - 41.25$
$= 19$ goals

These results can now be shown in a box plot.

0 20 40 60 80 100
Goals scored

> Although the box plot summarises the data well, showing the extreme values as well as the median and quartiles, it does **not** allow the original data to be retrieved.

Back to Back Stem and Leaf Diagrams

Comparisons between two sets of data can usefully be made if a back to back stem and leaf diagram is drawn.

The diagram below shows a back to back stem and leaf diagram comparing the times, in seconds, taken to run 800 m by a group of girls and a group of similarly aged boys.

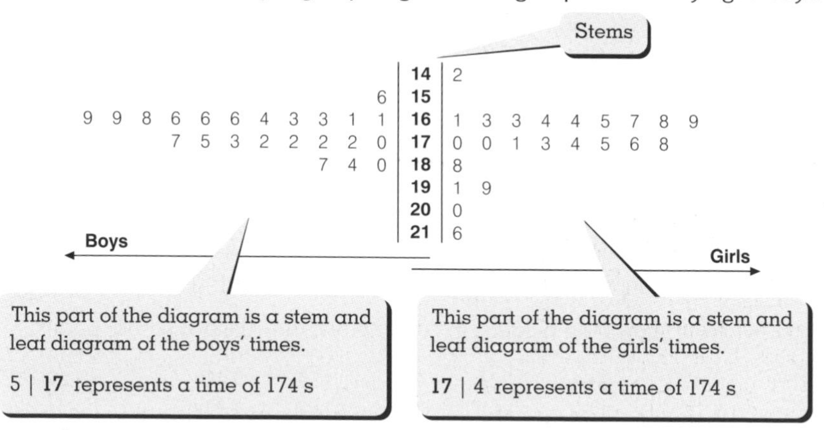

The diagram allows immediate comparisons to be made:

- there was a much smaller variability, or spread, in the times of the boys;
- the quickest athlete was a girl;
- the slowest athlete was also a girl.

Calculation of the medians and quartiles enables more precise comparisons to be made.

For the boys:

Median $= \frac{1}{2}(23 + 1)$
$= 12$th observation
$= 169$ s

Lower quartile $= \frac{1}{4}(23 + 1)$
$= 6$th observation
$= 164$ s

Upper quartile $= \frac{3}{4}(23 + 1)$
$= 18$th observation
$= 172$ s

Interquartile range $= 172 - 164 = 8$ s
Range $= 187 - 156 = 31$ s

For the girls:

Median $= \frac{1}{2}(23 + 1)$
$= 12$th observation
$= 170$ s

Lower quartile $= \frac{1}{4}(23 + 1)$
$= 6$th observation
$= 164$ s

Upper quartile $= \frac{3}{4}(23 + 1)$
$= 18$th observation
$= 178$ s

Interquartile range $= 178 - 164 = 14$ s
Range $= 216 - 142 = 74$ s

These calculations confirm the great difference in the variability, or spread, of the two sets of data, but also suggest there is little real difference between the average (median) time of the two sets of athletes.

A box plot can be drawn to emphasise these comments.

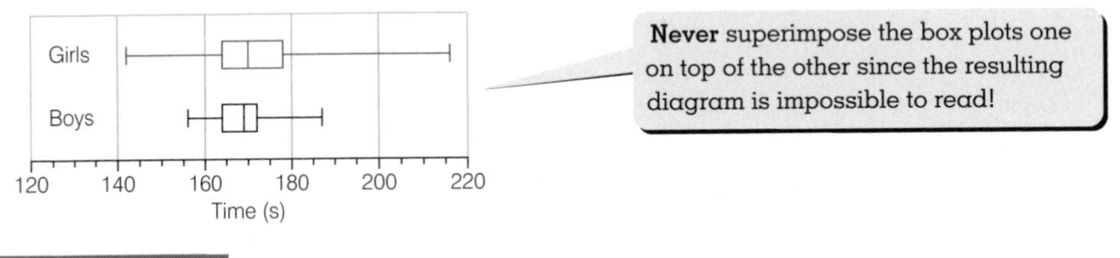

EXERCISE 2

1 Obtain the median and interquartile range of the data

a) 16 26 17 11 17 3 36

b) 126 124 127 105 128 124 115 126

c) 32 47 36 46 34 42 39 33 48

2 The table below shows the number of errors made by a typist on each of 40 different letters.

Number of errors	0	1	2	3	4	5	6
Frequency	6	14	13	3	2	1	1

Calculate the median and interquartile range of the data.

3 The table below shows the heights, in centimetres, of a class of school pupils.

161	171	165	157	149
153	163	177	182	171
151	175	184	148	159
168	170	167	155	164
160	181	163	157	168

a) Draw a stem and leaf diagram for the data.

b) Find the median and the quartiles for the data.

c) Draw a box diagram for the data.

The following diagram shows a box diagram for another class of 25 pupils in the school.

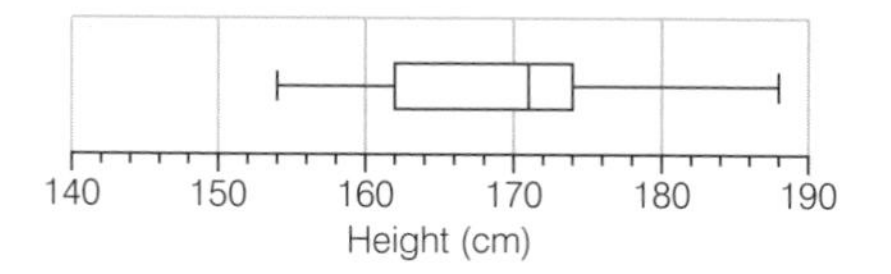

Write a short paragraph comparing the heights of the pupils in the two classes.

4 The data below shows the amount spent (to the nearest pound) by 25 customers passing through a cash till at a supermarket on a Wednesday afternoon.

27 36 21 9 23 47 31 25 18 4 13 20 29 30 21
17 24 32 18 56 37 26 32 15 27

a) Represent the data in a stem and leaf diagram.
b) Determine the median and interquartile range of the data. Represent the data with a box and whisker diagram.

The diagram below shows a box and whisker diagram for the amount spent by 25 customers passing through a cash till at the same supermarket on a Friday afternoon.

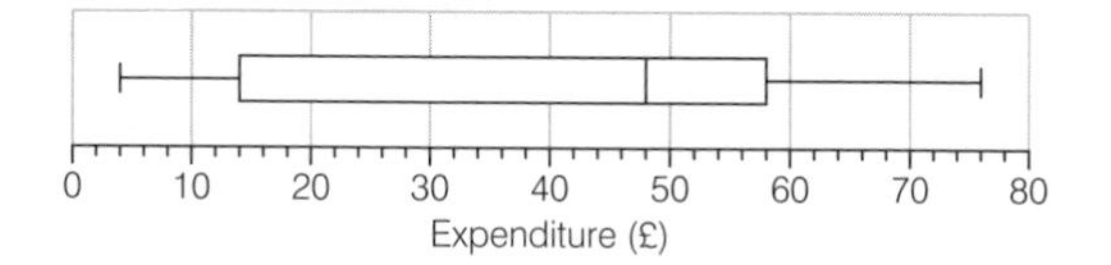

c) By comparing the two box and whisker diagrams, write a short paragraph comparing the expenditure of customers on a Wednesday and on a Friday.

5 The lengths, in seconds, of 20 calls made from a mobile phone on a "pay as you go" contract and the lengths of 20 calls made on a mobile phone on a "monthly rental" contract were sampled. The results are shown in the back to back stem and leaf diagram shown below.

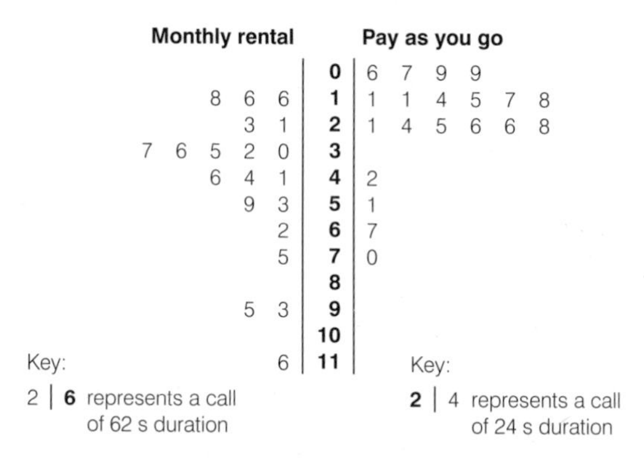

Monthly rental		Pay as you go
	0	6 7 9 9
8 6 6	**1**	1 1 4 5 7 8
3 1	**2**	1 4 5 6 6 8
7 6 5 2 0	**3**	
6 4 1	**4**	2
9 3	**5**	1
2	**6**	7
5	**7**	0
	8	
5 3	**9**	
	10	
6	**11**	

Key: 2 | **6** represents a call of 62 s duration

Key: **2** | 4 represents a call of 24 s duration

a) Determine the median and quartiles of the length of calls from the "monthly rental" phone.

The lower quartile, median and upper quartile lengths of calls for the "pay as you go" phone are 11 s, 19.5 s and 27.5 s, respectively.
b) Draw box plots to illustrate the two distributions.
c) Write a short paragraph comparing the lengths of calls made from phones with the monthly rental contract with those made from phones with the "pay as you go" contract.

Measures of Spread About the Mean

Consider the set of data given in the table below and illustrated by the vertical line diagram.

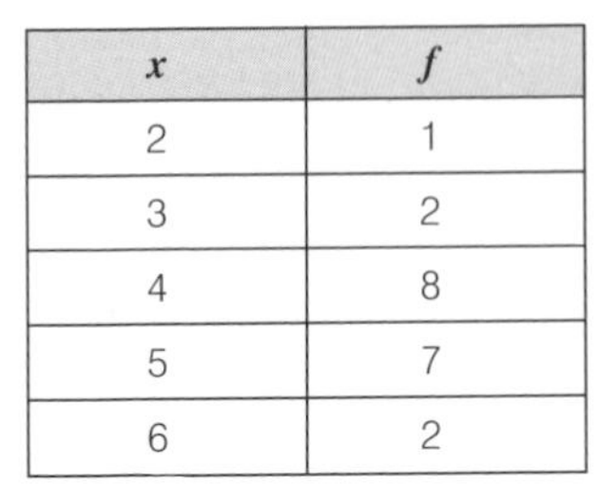

x	f
2	1
3	2
4	8
5	7
6	2

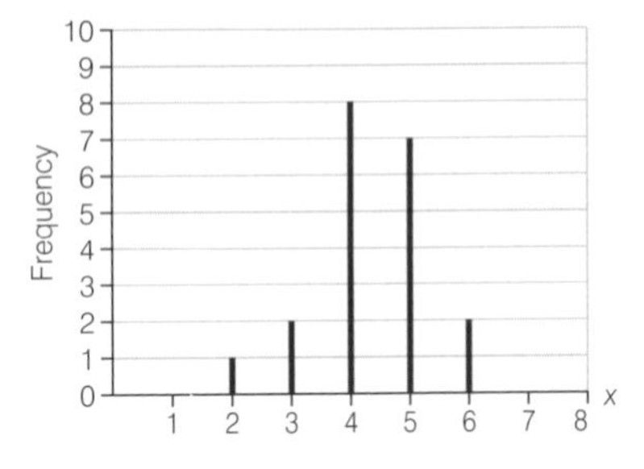

The mean of this data can easily be calculated:

x	f	xf
2	1	2
3	2	6
4	8	32
5	7	35
6	2	12
	20	87

$$\bar{x} = \frac{\sum xf}{\sum f} = \frac{87}{20} = 4.35$$

A measure of spread of the data can be evaluated by:

- calculating the deviation, d, of each piece of data away from the mean;
- then calculating the mean, $\bar{d}$, of these deviations.

The smaller the value of $\bar{d}$ the more closely packed the data are about the mean.

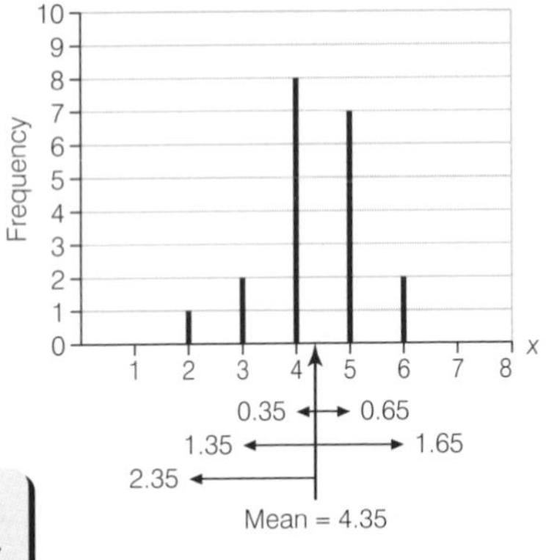

You know the mean is 4.35, so when

$x = 2$; deviation = 2.35
$x = 3$; deviation = 1.35
$x = 4$; deviation = 0.35
$x = 5$; deviation = 0.65
$x = 6$; deviation = 1.65

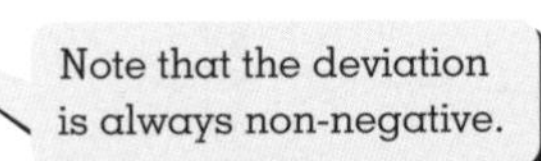

You can now work out the average of these deviations by returning to the frequency table.

x	f	xf	d	df
2	1	2	2.35	2.35
3	2	6	1.35	2.7
4	8	32	0.35	2.8
5	7	35	0.65	4.55
6	2	12	1.65	3.3
	20	87		15.7

The table now gives

$$\bar{d} = \frac{\text{total of } df \text{ column}}{\text{total frequency}} = \frac{\Sigma df}{\Sigma f} = \frac{15.7}{20} = 0.785.$$

This measure of spread is known as the mean absolute deviation from the mean, and it is a very natural way of measuring spread. It was much used in the early days of Statistics. However, its importance has declined during the last 150 years since it is quite difficult to establish the mathematical properties of this measure of spread.

This difficulty arises from the fact that the mathematical formulae for evaluating d are rather complex. Since d must be positive it is possible to write a two-part formula to express d:

if x is greater than the mean, $d = x - \bar{x}$
if x is less than the mean, $d = \bar{x} - x$

Alternatively, d can be written as the rather untidy formula

$d = \sqrt{(x - \bar{x})^2}$.

You will not need to calculate a mean absolute deviation from the mean since it has been completely replaced by the standard deviation as a measure of spread.

Variance and Standard Deviation

The measures of spread that have become by far the most commonly used are the **variance** and the **standard deviation**. These have the advantage of neat mathematical formulae. This enables easy evaluation with scientific calculators or computers. The mathematical properties of these measures can also be easily deduced with a modest amount of algebraic manipulation.

As an indicator of how far an observation is from the centre of the distribution, calculate the square of the deviation from the mean. Call it d^*. Then find the average (mean) of these values. This gives a measure of spread that can be used to compare distributions.

This measure of spread is called the **variance** of the data, and it can be calculated using the following steps:

- evaluate the mean, $\bar{x}$, of the data;
- work out $d^* = (x - \bar{x})^2$ for each piece of data;
- find the mean of the d^* values.

It is possible to summarise these steps in the formula

$$\text{Variance} = \frac{\Sigma d^* f}{N} = \frac{\Sigma (x - \bar{x})^2 f}{N} \text{ where } N = \Sigma f$$

EXAMPLE 7

Calculate the variance of the data given in the table.

x	f
2	1
3	2
4	8
5	7
6	2

EXAMPLE 7 (continued)

SOLUTION

The details of the calculation can be presented in a table.

x	f	xf	$d^* = (x - x)^2$	$d^*f = (x - \bar{x})^2 f$
2	1	2	5.5225	5.5225
3	2	6	1.8225	3.645
4	8	32	0.1225	0.98
5	7	35	0.4225	2.9575
6	2	12	2.7225	5.445
	20	87		18.55

$$\bar{x} = \frac{\Sigma xf}{\Sigma f} = \frac{87}{20} = 4.35$$

$$\text{Variance} = \frac{\Sigma d^* f}{N} = \frac{\Sigma (x - \bar{x})^2 f}{N} = \frac{18.55}{20} = 0.9275$$

For grouped data, the midmarks are used in the evaluation of the variance.

EXAMPLE 8

The times taken to complete a piece of homework by a class of 25 pupils are recorded in the table below. Calculate the variance of these times.

Time (mins)	10–20	20–30	30–40	40–50	50–60
Frequency	3	7	11	2	2

SOLUTION

Time (mins)	f	**Midmark, x**	xf	$d^* = (x - x)^2$	$d^*f = (x - \bar{x})^2 f$
10–20	3	15	45	295.84	887.52
20–30	7	25	175	51.84	362.88
30–40	11	35	385	7.84	86.24
40–50	2	45	90	163.84	327.68
50–60	2	55	110	519.84	1039.68
	25		805		2704

$$\bar{x} = \frac{\Sigma xf}{N} = \frac{805}{25} = 32.2 \text{ mins}$$

and

$$\text{Variance} = \frac{\Sigma d^* f}{N} = \frac{\Sigma (x - \bar{x})^2 f}{N} = \frac{2704}{25} = 108.16 \text{ mins}^2$$

Notice that, in this case, the data is measured in minutes but the variance, because of the squaring operation, is measured in mins^2.

In order to get a measure of spread which has the same units as the original data, it is usual to square root the variance to obtain the measure of spread known as the **standard deviation**.

Thus

Standard deviation of homework times = $\sqrt{\text{Variance}} = \sqrt{108.16} = 10.4$ min

In general, we can write

$$\text{Standard deviation} = \sqrt{\frac{\Sigma (x-\bar{x})^2 f}{N}} \text{ where } N = \Sigma f.$$

For the special case where the observations $x_1, x_2, \ldots, x_n$ each occur with frequency 1, the formulae for variance and standard deviation can be rewritten as

$$\text{Variance} = \frac{\Sigma (x-\bar{x})^2}{n}$$

$$\text{Standard deviation} = \sqrt{\frac{\Sigma (x-\bar{x})^2}{n}}$$

Interpretation of Standard Deviation

A useful result is the 2-standard deviation rule:

Most (at least 75% and often more than 95%) of any frequency distribution will lie in the interval between

mean – 2 × standard deviation and **mean + 2 × standard deviation**

Looking back at the last example, we can calculate that

mean – 2 × standard deviation = 32.2 – 2 × 10.4 = 11.4
mean + 2 × standard deviation = 32.2 + 2 × 10.4 = 53

so the 2-standard deviation rule tells us that most of the data should lie between 11.4 and 54 minutes.

If we now refer back to the frequency table, we can see that nearly all of the pupils did indeed spend between 11.4 and 53 minutes on their work.

EXERCISE 3

1 The table shows the number of absences of the pupils in a class during a particular term.

Number of days absent	0	1	2	3	4	5	6	7	8
Number of pupils	5	8	3	5	2	0	1	1	0

Calculate the mean, variance and standard deviation of this data.

2 The table shows the time spent by a commuter on his daily journey to work over a period of 50 days.

Time (mins)	40–	50–	60–	70–	80–90
Frequency	15	25	6	3	1

Find the mean, variance and standard deviation of these times.

3 Calculate the mean and standard deviation of the data

a) 42 36 27 38 38 26 27 41
b) 0 –7 1 –10 4

4 The data below show the number of fish caught by an angler on his last 20 fishing trips:

3 3 3 3 2 7 1 5 0 3 3 3 5 3 5 4 2 1 5 5

Calculate the mean and standard deviation of this data.

Alternative Formats for the Formula for Variance and Standard Deviation

The method of calculating the variance and hence the standard deviation is rather tedious but fortunately it can be greatly simplified by using the alternative formulae:

$$\textbf{Variance} = \frac{\sum (x-\bar{x})^2 f}{N} = \frac{\sum x^2 f}{N} - \bar{x}^2$$

$$\textbf{Standard deviation} = \sqrt{\frac{\sum (x-\bar{x})^2 f}{N}} = \sqrt{\frac{\sum x^2 f}{N} - \bar{x}^2}$$

We first check that the alternative formulae give the same answer as the original formulae for the two examples already considered.

For the first case, we had $\bar{x} = 4.35$ and, using the original variance formula, we obtained

Variance = 0.9275

Using the second formula, the calculation becomes

x	f	xf	x^2f
2	1	2	4
3	2	6	18
4	8	32	128
5	7	35	175
6	2	12	72
	20	87	397

This gives

$$\bar{x} = \frac{\sum xf}{N} = \frac{87}{20} = 4.35$$

and the alternative formula for variance gives

$$\text{Variance} = \frac{\sum x^2 f}{N} - \bar{x}^2 = \frac{397}{20} - 4.35^2 = 0.9275$$

Thus, for this example, the two formulae for variance do give the same answer.

For the data of Example 8, we had $\bar{x} = 32.2$ mins and, using the original variance formula, we obtained.

Variance = 108.16 mins^2.

Using the second formula, the calculation becomes

Time (mins)	f	Midmark, x	xf	$x^2 f$
10–20	3	15	45	675
20–30	9	25	175	4375
30–40	4	35	385	13475
40–50	3	45	90	4050
50–60	6	55	110	6050
	25		805	28625

This gives

$$\bar{x} = \frac{\sum xf}{N} = \frac{875}{25} = 35 \text{ mins}$$

and the alternative formula for variance gives

$$\text{Variance} = \frac{\sum x^2 f}{N} - \bar{x}^2 = \frac{28625}{25} - 32.2^2 = 108.16 \text{ mins}^2$$

Again, it can be seen that the two formulae do appear to give the same answer for the variance of a set of data.

In the Extension at the end of this chapter you will find a formal proof that the two formulae for variance will always give the same answer.

Summarising the results so far, we have:

For a frequency distribution where observations $x_1, x_2, \ldots$ occur with frequencies $f_1, f_2, \ldots$, respectively

1) **Mean** $= \bar{x} = \frac{\sum xf}{N}$

2) **Variance** $= \frac{\sum (x-\bar{x})^2 f}{N} = \frac{\sum x^2 f}{N} - \bar{x}^2$

3) **Standard deviation** $= \sqrt{\text{Variance}} = \sqrt{\frac{\sum x^2 f}{N} - \bar{x}^2}$

where $N = \sum f$

If each of the frequencies is 1 then the results simplify to the following.

If the values $x_1, x_2, \ldots, x_n$ are each taken with frequency 1 then

1) **Mean** $= \bar{x} = \frac{\sum x}{n}$

2) **Variance** $= \frac{\sum (x-\bar{x})^2}{n} = \frac{\sum x^2}{n} - \bar{x}^2$

3) **Standard deviation** $= \sqrt{\text{Variance}} = \sqrt{\frac{\sum x^2}{n} - \bar{x}^2}$

You should be able to use your graphical or scientific calculator to evaluate the standard deviation of a set of data. As before, it is important that you give some details of your method!

EXAMPLE 9

Find the mean, variance and standard deviation of the data

144 254 178 104 180

SOLUTION

Using a calculator, we have

$$n = 5$$

$$\bar{x} = \frac{\sum x}{n} = \frac{860}{5} = 172$$

$$\text{Variance} = \frac{\sum x^2}{n} - \bar{x}^2 = \frac{160\,152}{5} - 172^2 = 2446.4$$

$$\text{sd} = \sqrt{2446.4} = 49.46 \quad \text{(to 2 d.p.)}$$

We will use "sd" as a shorthand for standard deviation.

EXAMPLE 10

The table below shows the ages, in completed years, of the audience at a cinema.

Calculate the mean and standard deviation of the data.

Age (years)	0–19	20–29	30–39	40–59	60–89
Frequency	65	72	81	73	24

SOLUTION

Remembering that the true group boundaries are 0–20, 20–30, 30–40, etc., we have

Midmarks: 10, 25, 35, 50, 75

So

$$\bar{x} = \frac{\sum xf}{\sum f} = \frac{10\,735}{315} = 34.08 \text{ years} \qquad (2 \text{ d.p.})$$

and

$$\text{sd} = \sqrt{\frac{\sum x^2 f}{\sum f} - \bar{x}^2} = \sqrt{\frac{46\,8225}{315} - \left(\frac{10\,735}{315}\right)^2} = 18.03 \text{ years} \qquad (2 \text{ d.p.})$$

Note that $\frac{\sum x^2 f}{N}$ and $\bar{x}^2$ will often be very similar sized numbers, so to obtain accurate answers for the variance and standard deviation it is important to use the most accurate answer possible for $\bar{x}$ and this is usually the fraction $\frac{\sum xf}{N}$.

EXAMPLE 11

A group of data is summarised by

$\sum f = 120,\ \sum xf = 996,\ \sum x^2 f = 8502$

Find the mean and standard deviation of the data.
What can be deduced about the original data?

SOLUTION

$$\bar{x} = \frac{\sum xf}{N} = \frac{996}{120} = 8.3$$

$$\text{sd} = \sqrt{\frac{\sum x^2 f}{N} - \bar{x}^2} = \sqrt{\frac{8502}{120} - 8.3^2} = \sqrt{1.96} = 1.4$$

Since

mean − 2 × standard deviation = 8.3 − 2 × 1.4 = 5.5

and

mean + 2 × standard deviation = 8.3 + 2 × 1.4 = 11.1,

the '2 standard deviation' rule implies that most (at least three quarters) of the original data lies between 5.5 and 11.1.

EXAMPLE 12

A sample of 23 observations gave $\sum u = 312$ and $\sum u^2 = 5760$.

Find the mean and variance of this data.

SOLUTION

$$\bar{u} = \frac{\sum u}{n} = \frac{312}{23} = 13.57 \qquad \text{(2 d.p.)}$$

$$\text{Variance} = \frac{\sum u^2}{n} - \bar{u}^2 = \frac{5760}{23} - \left(\frac{312}{23}\right)^2 = \frac{35\,136}{529} = 66.42 \qquad \text{(2 d.p.)}$$

If you had used 13.57 as the value of $\bar{u}$ you would have obtained a variance of 66.29 to two decimal places.

Remember that the **precise** value of the mean must be used in the formula for variance or standard deviation if inaccuracies are to be avoided!

EXAMPLE 13

A sample of 45 observations gave $\sum (v - \bar{v})^2 = 5132$.

Find the standard deviation of this data.

SOLUTION

$$\text{sd} = \sqrt{\frac{\sum (v - \bar{v})^2}{n}} = \sqrt{\frac{5132}{45}} = 10.68 \qquad \text{(2 d.p.)}$$

EXERCISE 4

1 A boy threw a die 25 times and obtained the following scores:

3, 5, 1, 6, 5   6, 1, 1, 6, 5   3, 2, 4, 1, 1   4, 3, 3, 6, 6   3, 5, 3, 4, 2

Prepare a frequency table for these scores and hence calculate the mean and standard deviation of the scores.

2 The heights of 100 men are measured, **to the nearest centimetre**, and the results shown in the table below. Evaluate the mean and standard deviation of the heights.

Height (cm)	163–167	168–172	173–177	178–182	183–187	188–192	193–197
Frequency	5	10	24	29	19	10	3

3 The table below shows the ages, in completed years, of people on a coach holiday.

a) Draw a histogram to illustrate the data.

b) Evaluate the mean and standard deviation of the data.

Age (years)	0–39	40–49	50–59	60–69	70–89
Frequency	4	4	14	15	12

4 The table below shows the times, to the nearest minute, of telephone calls to a particular business number.

Time (mins)	0–2	3–5	6–8	9–13	14–20
Frequency	17	34	21	18	15

a) Draw a histogram to illustrate the data.
b) Calculate the mean and standard deviation of the lengths of these phone calls.

5 The table below shows the distribution of ages, in completed years, of teachers at three schools.

Age (years)	20–29	30–39	40–49	50–59	60–69
Number of teachers at School A	15	20	18	14	5
Number of teachers at School B	3	17	13	5	1
Number of teachers at School C	7	13	20	20	5

By summarising the data in suitable ways, compare the age distributions of the teachers at the three schools.

6 A set of data is summarised by $\sum f = 80$, $\sum xf = 132$, $\sum (x - \bar{x})^2 f = 112$.
Calculate the mean and variance of the data.

7 A set of data is summarised by $\sum f = 180$, $\sum xf = 512$, $\sum x^2 f = 1612$.
Calculate the mean and standard deviation of the data.
What can be said about the original data?

8 A sample consists of 24 observations with $\sum x = 614$, $\sum x^2 = 15\,782$.
Calculate the mean and standard deviation of the data.
What can be said about the original data?

9 A sample consists of 40 observations with $\sum (x - \bar{x})^2 = 328$.
Calculate the standard deviation of the data.

10 **a)** Calculate the mean and variance of the following set of data:

x	0	1	2	3	4
f	2	5	9	3	1

b) A new set of data is obtained from this set using the rule $y = 3x + 2$. This gives

y	2	5	8	11	14
f	2	5	9	3	1

Calculate the mean and variance of this set of data.

c) A new set of data is obtained from this set using the rule $z = 5x + 1$. This gives

z	1	6	11	16	21
f	2	5	9	3	1

Calculate the mean and variance of this set of data.
What is the rule linking the mean of x to the means of y and z?
What is the rule linking the variance of x to the variances of y and z?

EXTENSION

Proof of the Equivalence of the Two Formulae for Variance

If observations $x_1, x_2, \ldots, x_n$ occur with frequencies $f_1, f_2, \ldots, f_n$ then

$$\begin{aligned}
\text{Variance} &= \frac{\sum (x-\bar{x})^2 f}{\sum f} \\
&= \frac{(x_1-\bar{x})^2 f_1 + (x_2-\bar{x})^2 f_2 + \cdots + (x_n-\bar{x})^2 f_n}{\sum f} \\
&= \frac{(x_1^2 - 2x_1\bar{x} + \bar{x}^2)f_1 + (x_2^2 - 2x_2\bar{x} + \bar{x}^2)f_2 + \cdots + (x_n^2 - 2x_n\bar{x} + \bar{x}^2)f_n}{\sum f} \\
&= \frac{(x_1^2 f_1 + x_2^2 f_2 + \cdots + x_n^2 f_n) - (2x_1\bar{x}f_1 + 2x_2\bar{x}f_2 + \cdots + 2x_n\bar{x}f_n) + (\bar{x}^2 f_1 + \bar{x}^2 f_2 + \cdots + \bar{x}^2 f_n)}{\sum f} \\
&= \frac{\sum x^2 f - 2\bar{x}(x_1 f_1 + x_2 f_2 + \cdots + x_n f_n) + \bar{x}^2(f_1 + f_2 + \cdots + f_n)}{\sum f} \\
&= \frac{\sum x^2 f - 2\bar{x}\sum xf + \bar{x}^2 \sum f}{\sum f} \\
&= \frac{\sum x^2 f}{\sum f} - 2\bar{x}\frac{\sum xf}{\sum f} + \bar{x}^2 \frac{\sum f}{\sum f}
\end{aligned}$$

but

$$\bar{x} = \frac{\sum xf}{\sum f} \quad \text{and} \quad \frac{\sum f}{\sum f} = 1$$

so

$$\begin{aligned}
\text{Variance} &= \frac{\sum x^2 f}{\sum f} - 2\bar{x}\bar{x} + \bar{x}^2 \times 1 \\
&= \frac{\sum x^2 f}{\sum f} - 2\bar{x}^2 + \bar{x}^2 \\
&= \frac{\sum x^2 f}{\sum f} - \bar{x}^2
\end{aligned}$$

as required.

Having studied this chapter you should know how to

- find the range and interquartile range of a set of data
- draw and interpret box and whisker diagrams
- draw and interpret stem and leaf diagrams
- calculate the mean, variance and standard deviation of observations $x_1, x_2, \ldots$ occurring with frequencies $f_1, f_2, \ldots$, respectively using the formulae

Mean $= \bar{x} = \dfrac{\Sigma xf}{N}$

Variance $= \dfrac{\Sigma (x-\bar{x})^2 f}{N} = \dfrac{\Sigma x^2 f}{N} - \bar{x}^2$

Standard deviation $= \sqrt{\text{Variance}} = \sqrt{\dfrac{\Sigma x^2 f}{N} - \bar{x}^2}$

where $N = \Sigma f$

- calculate the mean, variance and standard deviation of observations $x_1, x_2, \ldots, x_n$ occurring with frequency 1 using the formulae

Mean $= \bar{x} = \dfrac{\Sigma x}{n}$

Variance $= \dfrac{\Sigma (x-\bar{x})^2}{n} = \dfrac{\Sigma x^2}{n} - \bar{x}^2$

Standard deviation $= \sqrt{\text{Variance}} = \sqrt{\dfrac{\Sigma x^2}{n} - \bar{x}^2}$

- use "the 2-standard deviation rule" to make deductions about a distribution from knowledge of its mean and standard deviation

REVISION EXERCISE

1 The table below shows the journey times, in minutes to the nearest minute, of trains travelling between London Paddington and Reading on one day.

Time (mins)	30–39	40–44	45–49	50–59	60–74
Number of trains	24	21	17	12	6

a) Draw a histogram to represent this data.
b) Calculate the mean and standard deviation of this set of data.

2 The diagram below shows a box and whisker plot comparing the marks of a class of pupils in a Maths exam and in an English exam.

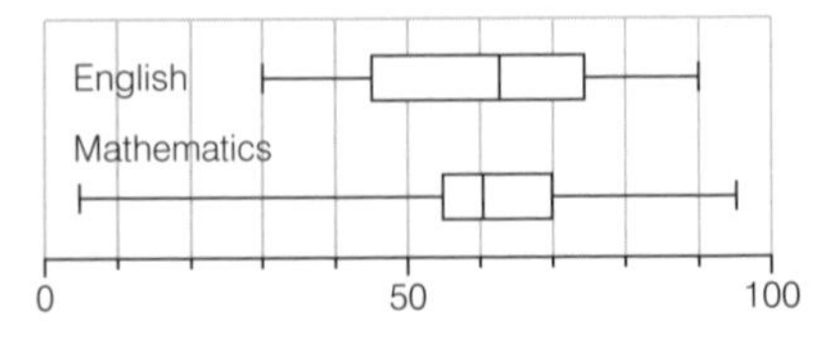

Write a paragraph comparing the two sets of marks.

3 The distances travelled, correct to the nearest 10 km, of the people attending a company's 2003 Annual General Meeting in London are summarised in the table below.

Distance (km)	0–20	30–50	60–100	110–200	210–400
Frequency	32	43	37	42	26

a) Draw a histogram to illustrate the data.
b) Calculate, showing your method clearly, the mean and standard deviation of the data.

In 2004 the company held its AGM in Leeds. One hundred and forty five people attended the meeting; their mean travelling distance was 175 km and the standard deviation of the travelling distances was 62.4 km.

Write a paragraph comparing the travelling distances of people attending the 2003 and the 2004 AGMs.

4 The table below shows the length of films shown at a cinema during the course of a year.

Time (mins)	60–100	100–110	110–120	120–150	150–180
Frequency	7	9	13	8	4

a) Draw a cumulative frequency graph for the data and hence find the median and interquartile range of the data.
b) Draw a box plot to illustrate the data.
c) Find the percentage of the films that lasted between 108 and 130 minutes.

5 A senior officer at a fire station carried out an investigation into the time taken for a fire engine to reach the scene of an emergency. He recorded the time, t minutes, between receiving a call for assistance and the arrival of the engine at the scene of the emergency. The results are given in the table below.

t	**Frequency**
$0 < t \leqslant 2$	6
$2 < t \leqslant 4$	8
$4 < t \leqslant 6$	7
$6 < t \leqslant 10$	9
$10 < t < 20$	20

i) On graph paper, draw a cumulative frequency graph to represent the data.
ii) From your graph estimate:
  a) the median time;
  b) the upper quartile of the times.

(OCR Nov 2002 S1)

6 The stem and leaf diagram below shows the resting pulse beat, in heartbeats per minute, for a sample of 34 female students and 14 male students.

Female students		Male students
9 4 4	**5**	
8 8 7 7 7 7 6 6 4 2	**6**	5 6 7
9 7 7 6 6 5 4 2 2 1 0 0	**7**	1 2 5 5 5 6 9
9 8 8 4 4 3 2 0 0	**8**	3 3
	9	5 7

For example, in the second row of the diagram, 2 | **6** | 5 *represents a female student's resting pulse rate of 62 and a male student's resting pulse rate of 65.*

i) Calculate:

a) the median of the male students' resting pulse rates;

b) the interquartile range of the male students' resting pulse rates.

ii) On graph paper draw a box and whisker plot to represent the data for the male students.

The median resting pulse rate for the female students is 72. The lower and upper quartiles of the resting pulse rates for the female students are 67 and 80, respectively.

iii) On the same diagram as in part (b), draw a box and whisker plot to represent the data for female students.

iv) Using your two box and whisker plots, state one difference between the data for the male students and the data for the female students.

(OCR Jan 2001 S1)

7 A set of data is summarised by $\Sigma f = 120$, $\Sigma xf = 232$, $\Sigma x^2 f = 510$.

Calculate the mean and standard deviation of the data. What can be deduced about the original data from these calculations?

8 A consumer organisation conducted a trial to investigate the durability of tyres of a particular make. Two hundred of the tyres were fitted to the rear wheels of cars of the same model and the distances travelled by the tyres before reaching the legal limit of tyre wear were recorded. The results are summarised in the table below.

Distance (x km)	Number of tyres
$15\,000 < x \leqslant 20\,000$	20
$20\,000 < x \leqslant 30\,000$	46
$30\,000 < x \leqslant 40\,000$	62
$40\,000 < x \leqslant 50\,000$	38
$50\,000 < x \leqslant 60\,000$	24
$60\,000 < x \leqslant 80\,000$	10

a) Draw a histogram to illustrate this distribution.

b) Calculate, showing your working, estimates for the mean and standard deviation of the distribution.

The manufacturer of the tyres decides to conduct a simulated tyre wear trial by running 200 tyres on constant speed rollers and then calculate the mean and standard deviation of the simulated distances travelled by the tyres before reaching the legal limit of tyre wear.

c) Comment briefly upon the difference you would expect between:

i) the means obtained in the two trials;

ii) the standard deviations obtained in the two trials.

9 Two chess players, A and B, play a tournament. Each player plays against 11 different opponents. Each game concludes when either the player has won, his opponent has won, or the game has ended in a draw. The time, correct to the nearest minute, for each game to conclude was noted and the results are given below.

Player A:	7	23	32	32	33	41	46	56	56	61	62
Player B:	3	7	8	8	14	26	27	37	37	41	58

i) For player A's times find:
a) the median;
b) the upper quartile;
c) the lower quartile.

ii) On graph paper, using the same axis, draw two box and whisker plots to represent the times of the games played by A and B.

iii) Use your diagram to make two statements comparing the two sets of data.

iv) Can you state from the diagram or from the data which player was the more successful? Give a reason for your answer.

(OCR May 2002 S1)

10 As part of a statistics project a student recorded the amount of money spent, in pounds, by each of a random sample of 60 customers at checkout A in a supermarket. She also recorded the amount spent by each of a random sample of 60 customers who used another checkout B in the same supermarket. The results are given in the table below.

Amount spent	**⩽£10**	**⩽£20**	**⩽£40**	**⩽£60**	**⩽£100**
Cumulative frequency for checkout A	25	41	52	56	60
Cumulative frequency for checkout B	10	24	45	54	60

The diagram shows the cumulative frequency graph for the data.

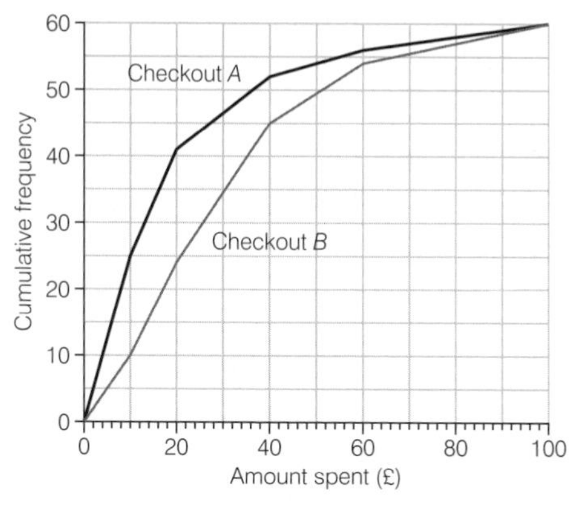

i) Use the diagram to estimate the median amount spent at:
a) checkout A; **b)** checkout B.

ii) Use the diagram to estimate the interquartile range of the amount spent at:
a) checkout A; **b)** checkout B.

iii) One of the two checkouts was "express" checkout. Customers are allowed a maximum of nine items when they pass through an express checkout. State, with a reason, which of the two checkouts, A or B, was more likely to have been the express checkout.

iv) Calculate an estimate of the mean amount spent at checkout B.

(OCR Jun 2001 S1)

3 Working with Mean and Standard Deviation

The purpose of this chapter is to enable you to

- calculate the mean, variance and standard deviation of data linked by a linear relationship to data whose mean, variance and standard deviation are already known
- find the mean and standard deviation of a set of data formed by combining two separate sets of data whose means and standard deviations are known

Coding: The Mean, Variance and Standard Deviation of $y = ax + b$

Two frequency distributions may often be linked by a simple linear rule. Consider the distributions

x	2	3	4	5
f	5	12	7	6

and

y	1	4	7	10
f	5	12	7	6

They have **exactly the same frequencies** and the y values in the second distribution are linked to the x values of the first distribution by the rule $y = 3x - 5$.

In this section a quick method of calculating the mean, variance and standard deviation of the second distribution from the respective values of the first distribution will be determined.

EXAMPLE 1

a) The table below shows a frequency distribution. Calculate the mean, variance and standard deviation of the distribution.

x	0	1	2	3	4
f	3	8	5	3	1

b) New data values, y, are to be calculated from these x values using the rule $y = 2x + 5$. Produce the frequency distribution for the y values and calculate the mean, variance and standard deviation of the y values.

SOLUTION

a) Using a calculator produces

$$\bar{x} = \frac{\Sigma xf}{N} = \frac{31}{20} = 1.55$$

EXAMPLE 1 (continued)

$$\text{Variance} = \frac{\Sigma x^2 f}{N} - \bar{x}^2 = \frac{71}{20} - (1.55)^2 = 1.1475$$

$$\text{Standard deviation} = \sqrt{\text{Variance}} = \sqrt{1.1475} = 1.071 \qquad \text{(to 3 d.p.)}$$

b) The new frequency distribution is

x	0	1	2	3	4
$y = 2x + 5$	5	7	9	11	13
f	3	8	5	3	1

and using a calculator produces

$$\bar{y} = \frac{\Sigma yf}{N} = \frac{162}{20} = 8.1$$

$$\text{Variance} = \frac{\Sigma y^2 f}{N} - \bar{y}^2 = \frac{1404}{20} - (8.1)^2 = 4.59$$

$$\text{Standard deviation} = \sqrt{\text{Variance}} = \sqrt{4.59} = 2.142 \qquad \text{(to 3 d.p.)}$$

Notice that

$$8.1 = \bar{y} = 2 \times 1.55 + 5 = 2\bar{x} + 5$$

or, in other words, the value of $\bar{y}$ is linked to the value of $\bar{x}$ **by the same linear rule** that gives the y values.

Notice also that

$$4.59 = \text{variance of } y \text{ values} = 4 \times 1.1475 = 2^2 \times \text{variance of } x \text{ values}$$

and that

$$2.142 \ldots = \text{standard deviation of } y \text{ values} = 2 \times 1.071 \ldots$$
$$= 2 \times \text{standard deviation of } x \text{ values}$$

These results can be generalised:

Suppose there are data values $x_1, x_2, x_3, \ldots, x_n$, occurring with frequencies $f_1, f_2, f_3, \ldots, f_n$, and new data values $y_1, y_2, y_3, \ldots, y_n$, are obtained from the old data values by using the linear rule

$$y = ax + b$$

where a is a positive constant and b is a constant. Then

$\bar{y} = a\bar{x} + b$

variance of y = $a^2 \times$ variance of x

standard deviation of y = $a \times$ standard deviation of x

A proof of this important result is presented in the Extension to this chapter.

EXAMPLE 2

The data below shows the number of haircuts that a barber does each hour during a 40-hour week.

Number of haircuts (x)	0	1	2	3	4	5	6
Frequency	3	5	12	11	7	1	1

a) Calculate the mean and standard deviation of the number of haircuts per hour.

b) The barber charges £9 for each haircut. His overheads amount to £20 per hour. If y denotes the profit he makes in an hour when he does x haircuts, obtain a relationship linking y to x and hence find the mean and standard deviation of y.

SOLUTION

a) Using a calculator produces

$$\bar{x} = \frac{\sum xf}{N} = \frac{101}{40} = 2.525 \text{ haircuts}$$

$$\text{Variance of } x = \frac{\sum x^2 f}{N} - \bar{x}^2 = \frac{325}{40} - 2.525^2 = 1.749.......$$

$$\text{Standard deviation of } x = \sqrt{1.749.......} = 1.323 \text{ haircuts} \qquad \text{(to 3 d.p.)}$$

b) If the barber does x haircuts in an hour then his income is $9x$. His profit, y, is his income less his overheads, which amount to £20, so

$$y = 9x - 20$$

$$\Rightarrow \quad \bar{y} = 9 \times 2.525 - 20 = 2.725$$

and

$$\text{sd } y = 9 \times \text{sd } x = £11.90 \qquad \text{(to 2 d.p.)}$$

EXAMPLE 3

A sample of 50 pieces of data gave the following results:

$$\sum x = 320 \qquad \sum x^2 = 2248$$

a) Calculate the mean and standard deviation of the data.

b) If new data values are obtained from the x values using the rule $y = 2x - 5$, calculate the mean and standard deviation of the new values.

SOLUTION

a) $\bar{x} = \frac{\sum x}{n} = \frac{320}{50} = 6.4 \qquad \text{sd}x = \sqrt{\frac{\sum x^2}{n} - \bar{x}^2} = \sqrt{\frac{2248}{50} - 6.4^2} = 2$

b) Since $y = 2x - 5$

$\bar{y} = 2\bar{x} - 5 = 2 \times 6.4 - 5 = 7.8$

and $\text{sd } y = 2 \times \text{sd } x = 4$

EXAMPLE 4

A sample of 80 pieces of data gave the following results:

$$\sum x = 560 \qquad \sum (x - \bar{x})^2 = 720$$

a) Calculate the mean and standard deviation of the data.
b) If new data values are obtained from the x values using the rule $y = 4x + 5$, calculate the mean and standard deviation of the new values.

SOLUTION

a) $$\bar{x} = \frac{\sum x}{n} = \frac{560}{80} = 7 \qquad \text{sd } x = \sqrt{\frac{\sum (x - \bar{x})^2}{n}} = \sqrt{\frac{720}{80}} = 3$$

b) $$y = 4x + 5$$

$$\Rightarrow \quad \bar{y} = 4\bar{x} + 5 = 4 \times 7 + 5 = 33$$

and

$$\text{sd } y = 4 \times \text{sd } x = 12.$$

EXAMPLE 5

A sample of 120 pieces of data gave the following results:

$$\sum (x - 5) = 300 \qquad \sum (x - 5)^2 = 840$$

a) Calculate the mean and standard deviation of the data.
b) Find also the values of $\sum (2x - 13)$ and $\sum (2x - 13)^2$.

SOLUTION

a) Let $\quad y = x - 5$

Then $\quad \sum y = 300 \Rightarrow \bar{y} = \frac{\sum y}{n} = \frac{300}{120} = 2.5$

and $\quad \sum y^2 = 840 \Rightarrow \text{sd } y = \sqrt{\frac{\sum y^2}{n} - \bar{y}^2} = \sqrt{\frac{840}{120} - 2.5^2} = \sqrt{\frac{3}{4}} = \frac{\sqrt{3}}{2}$

Now $\quad x = y + 5 \Rightarrow \bar{x} = \bar{y} + 5 = 7.5 \quad$ and $\quad \text{sd } x = 1 \times \text{sd } y = \frac{\sqrt{3}}{2}$

b) Let $\quad z = 2x - 13$

Then $\quad \bar{z} = 2\bar{x} - 13 = 2 \times 7.5 - 13 = 2$

So $\quad \bar{z} = \frac{\sum z}{120} = 2$

$$\Rightarrow \quad \sum z = 240$$

$$\Rightarrow \quad \sum (2x - 13) = 240$$

and $\quad$ Variance of $z = 2^2 \times$ Variance of $x = 4 \times \left(\frac{\sqrt{3}}{2}\right)^2 = 3$

so $\quad$ Variance of $z = \frac{\sum z^2}{120} - \bar{z}^2 = 3$

$$\Rightarrow \quad \frac{\sum z^2}{120} - 2^2 = 3$$

$$\Rightarrow \quad \sum z^2 = 840$$

$$\Rightarrow \quad \sum (2x - 13)^2 = 840$$

Historically, the coding result was of great importance since it gave an efficient way of evaluating the mean, variance and standard deviation of many sets of data even when a calculator was not available. The following example illustrates this.

EXAMPLE 6

The data below shows the heights, to the nearest centimetre, of tree saplings after 12 months growth in a nursery. Without using a calculator, calculate the mean and standard deviation of this data.

Height (cm)	100–109	110–119	120–129	130–139	140–149
Frequency	3	12	5	4	1

SOLUTION

The midmarks (x) of the groups are 104.5, 114.5, 124.5, 134.5, 144.5.

Direct evaluation of the mean and variance would require calculation of quantities such as 104.5×3 and $104.5^2 \times 3$, which would be quite difficult without a calculator.

The alternative process is **coding the data** using the rule $y = \dfrac{x - 104.5}{10} = 0.1x - 10.45$, which turns the midmarks 104.5, 114.5, 124.5, 134.5, 144.5 into y values 0, 1, 2, 3, 4. It is then possible to calculate the mean and variance of the y values and then deduce the mean and variance of the x values.

Height (cm)	**Frequency** f	**Midmark** x	y	yf	y^2f
99.5–109.5	3	104.5	0	0	0
109.5–119.5	12	114.5	1	12	12
119.5–129.5	5	124.5	2	10	20
129.5–139.5	4	134.5	3	12	36
139.5–149.5	1	144.5	4	4	16
	25			38	84

$$\bar{y} = \frac{\Sigma yf}{N} = \frac{38}{25} = 1.52$$

$$\text{Variance of } y = \frac{\Sigma y^2 f}{N} - \bar{y}^2 = \frac{84}{25} - 1.52^2 = 3.36 - 2.3104 = 1.0496$$

$$\text{Standard deviation of } y = \sqrt{1.0496} = 1.024 \qquad \text{(3 d.p.)}$$

The square root would have been found from a book of tables in the days before calculators!

Since

$$y = \frac{x - 104.5}{10} \Rightarrow x = 10y + 104.5$$

the coding results give

$$\bar{x} = 10\bar{y} + 104.5 = 15.2 + 104.5 = 119.7 \text{ cm}$$

and

$$\text{standard deviation of } x = 10 \times \text{standard deviation of } y = 10.24\text{cm} \qquad \text{(2 d.p.)}$$

EXERCISE 1

1 The frequency distribution below shows the number, x, of cars sold by a small car showroom each week over the period of a year.

x	3	4	5	6	7	8	9	10	11	12
f	2	3	6	6	8	11	9	5	1	1

a) Calculate the mean, variance and standard deviation of the data.

The showroom makes a profit of £2100 on each car sold but has overheads amounting to £4700 per week.

b) Find an expression for the profit £y made by the showroom when it sells x cars in a week.

c) Calculate the mean, variance and standard deviation of the weekly profit made by the showroom.

2 A class of 50 students has sat a multiple choice test of 10 questions. The table below shows the number of correct responses obtained by members of the class.

No. of correct responses	3	4	5	6	7	8	9	10
Frequency	2	5	8	17	10	4	3	1

a) Calculate the mean and standard deviation of the number of correct responses obtained by these pupils.

For each correct response the student scores 4 marks. For each incorrect response or question not answered the student scores –1 mark.

b) Suppose a student gets x correct responses. Show that his resulting score, y, is given by the formula $y = 5x - 10$.

c) Hence determine the mean and standard deviation of the pupils' scores on this test.

3 A sample of 30 pieces of data gave the following results:

$$\sum x = 540 \qquad \sum x^2 = 10\,470$$

a) Calculate the mean and standard deviation of the data.

b) If new data values are obtained from the x values using the rule $y = 6x + 8$, calculate the mean and standard deviation of the new values.

4 A sample of 60 pieces of data gave the following results:

$$\sum x = 72 \qquad \sum x^2 = 91.8$$

a) Calculate the mean and standard deviation of the data.

b) If new data values are obtained from the x values using the rule $y = 23 + 10x$, calculate the mean and standard deviation of the new values.

5 A sample of 40 pieces of data gave the following results:

$$\sum (x - 8) = 300 \qquad \sum (x - 8)^2 = 2410$$

a) Calculate the mean and standard deviation of the data.

b) Find also the values of $\sum (x - 12)$ and $\sum (x - 12)^2$.

6 A sample of 20 pieces of data gave the following results:

$$\sum(x+3) = 100 \qquad \sum(x+3)^2 = 503.2$$

Calculate the mean and standard deviation of the data.

7 A sample of 50 pieces of data gave

$$\sum(3x-5) = 225 \qquad \sum(3x-5)^2 = 1125$$

Calculate the mean and standard deviation of the data.

Finding the Mean and Standard Deviation of a Combined Sample

There will be times when the mean and standard deviation of two separate sets of data have been calculated and it is necessary to combine the results to obtain a mean and standard deviation for the combined set of data. This can be done provided the size of each set of data is known as well as the means and standard deviations.

EXAMPLE 7

A sixth form college has 80 girls and 60 boys in the first year.

The mean number of GCSE grades A–C for the girls was 7.2 and the standard deviation was 2.6.

The mean number of GCSE grades A–C for the boys was 6.5 and the standard deviation was 3.5.

Find the mean and the standard deviation of the number of GCSE grades A–C obtained by the pupils of the whole year.

SOLUTION

First consider the means.

For the girls:

$$7.2 = \overline{x_G} = \frac{(\sum xf)_G}{80} \Rightarrow (\sum xf)_G = 80 \times 7.2 = 576$$

This is the total number of GCSE grades A–C for the girls.

For the boys:

$$6.5 = \overline{x_B} = \frac{(\sum xf)_B}{60} \Rightarrow (\sum xf)_B = 60 \times 6.5 = 390$$

This is the total number of GCSE grades A–C for the boys.

For the pupils in the whole year:

$$\bar{x} = \frac{\sum xf}{N} = \frac{(\sum xf)_G + (\sum xf)_B}{140} = \frac{576 + 390}{140} = 6.9 \text{ passes/pupil}$$

Now consider the standard deviations.

For the girls, the standard deviation was 2.6 so the variance was $2.6^2 = 6.76$

$$\Rightarrow \quad 6.76 = \frac{(\sum x^2 f)_G}{80} - 7.2^2 \Rightarrow (\sum x^2 f)_G = (6.76 + 7.2^2) \times 80 = 4688$$

EXAMPLE 7 (continued)

For the boys, the standard deviation was 3.5 so the variance was $3.5^2 = 12.25$

$$\Rightarrow \qquad 12.25 = \frac{(\sum x^2 f)_B}{60} - 6.5^2 \Rightarrow (\sum x^2 f)_B = (12.25 + 6.5^2) \times 60 = 3270$$

For the pupils in the whole year:

$$\text{Variance} = \frac{\sum x^2 f}{N} - \bar{x}^2 = \frac{(\sum x^2 f)_G + (\sum x^2 f)_B}{140} - 6.9^2$$

$$\Rightarrow \qquad \text{Variance} = \frac{4688 + 3270}{140} - 6.9^2 = 9.232 \ldots$$

$$\Rightarrow \qquad \text{Standard deviation} = \sqrt{9.232 \ldots} = 3.04 \text{ passes} \qquad (2 \text{ d.p.})$$

EXAMPLE 8

The distances $x_1, x_2, x_3, \ldots, x_{12}$, in kilometres, travelled to work by the 12 people who work in an office are summarised by

$$\sum (x - 30) = 42 \qquad \sum (x - 30)^2 = 654$$

a) Calculate the mean and standard deviation of these distances.
b) Obtain the value of $\sum x$ and $\sum x^2$.

A new employee joins the office staff, whose travelling distance to work is 53 km.

c) Find the mean and standard deviation of the travelling distances for the group of 13 employees.

SOLUTION

a) Let $\qquad y = x - 30$

then $\qquad \sum y = 42 \Rightarrow \bar{y} = \frac{\sum y}{n} = \frac{42}{12} = 3.5$

and $\qquad \sum y^2 = 654 \Rightarrow \text{sd } y = \sqrt{\frac{\sum y^2}{n} - \bar{y}^2} = \sqrt{\frac{654}{12} - 3.5^2} = \sqrt{42.25} = 6.5 \text{ km}$

Now $\qquad x = y + 30 \Rightarrow \bar{x} = \bar{y} + 30 = 33.5 \qquad$ and $\qquad \text{sd } x = 1 \times \text{sd } y = 6.5 \text{ km}$

b) It is known that

$$33.5 = \bar{x} = \frac{\sum x}{n} = \frac{\sum x}{12}$$

$$\Rightarrow \qquad \sum x = 12 \times 33.5 = 402 \text{ km}$$

It is also known that

$$6.5 = \text{sd} = \sqrt{\frac{\sum x^2}{n} - \bar{x}^2} = \sqrt{\frac{\sum x^2}{12} - 33.5^2}$$

$$\Rightarrow \qquad 6.5^2 = \frac{\sum x^2}{12} - 33.5^2$$

$$\Rightarrow \qquad 1164.5 = \frac{\sum x^2}{12}$$

$$\Rightarrow \qquad \sum x^2 = 13\,974$$

EXAMPLE 8 (continued)

c) For the enlarged group of 13 employees

$$\sum x = 402 + 53 = 455 \qquad \sum x^2 = 13\,974 + 53^2 = 16\,783$$

So

$$\bar{x} = \frac{\sum x}{n} = \frac{455}{13} = 35 \text{ km}$$

and

$$\text{sd} = \sqrt{\frac{\sum x^2}{n} - \bar{x}^2} = \sqrt{\frac{16\,783}{13} - 35^2} = \sqrt{66} = 8.12 \text{ km} \qquad (3 \text{ s.f.})$$

EXERCISE 2

1 A group of 20 male and 15 female athletes ran in a marathon race. The running times of the males had a mean of 164 minutes and the running times of the females had a mean of 185 minutes.

Calculate the mean running time for the group of 35 athletes.

2 At a nursing home there are 30 female and 20 male residents. The ages of the female residents, in completed years, are given below:

Age	60–69	70–74	75–79	80–84	85–99
Frequency	6	12	8	2	2

a) Draw a histogram to represent the data.
b) Calculate the mean and standard deviation of the ages of the female residents.

The mean age of the male residents is 74 and the standard deviation of their ages is 3.5 years.

c) Calculate the mean and standard deviation of the ages of all the residents of the nursing home.

3 The times $x_1, x_2, x_3, \ldots, x_{20}$, in minutes, taken by 20 employees to complete a task are summarised by

$$\sum (x - 50) = 140 \qquad \sum (x - 50)^2 = 4360$$

a) Calculate the mean and standard deviation of these times.
b) Obtain the value of $\sum x$ and prove that $\sum x^2 = 68\,360$.

Two more employees completed the task in times of 79 and 90 minutes.

c) Find the mean and standard deviation of the times for the group of 22 employees.

4 The annual incomes $v_1, v_2, \ldots, v_{50}$, measured in thousands of pounds, of the 50 office workers of a company are summarised by

$$\sum v = 1170 \qquad \sum v^2 = 28\,482.5$$

a) Obtain the mean and standard deviation of this data.

The company employs another 180 people at its factory. The mean salary of these people is £19 950 and the standard deviation of these salaries is £3800. Let $w_1, w_2, \ldots, w_{180}$ denote the annual incomes, in thousands of pounds, of these employees.

b) Prove that $\sum w = 3591$.

c) Prove that $\sum w^2 = 74\,239.65$.

d) Find the mean and standard deviation of the annual incomes of the company's 230 employees.

5 Each day express and slow trains travel between two cities. One day the travelling times of these trains were measured and the data collected are summarised in the table below.

	Number of trains	Mean travelling time (mins)	Standard deviation of travelling times (mins)
Express Trains	20	85	6
Slow Trains	30	92.5	10

Obtain the mean and standard deviation of the combined sample of 50 trains.

6 The masses of a sample of 20 male adult spiders and a sample of 16 female adult spiders of the same species were measured.

The masses of the male spiders had mean 32.5 grams and variance 25 grams^2, whilst the masses of the female spiders had mean 75 grams and variance 37.5 grams^2.

a) Calculate the mean mass of the combined sample of 36 spiders.

b) Prove that the variance of the combined sample of 36 spiders is 476.5 grams^2 correct to one decimal place.

The distributions of the masses are shown in the histogram below. Use this diagram to explain why the variance of the combined sample is much larger than the variances of each separate sample.

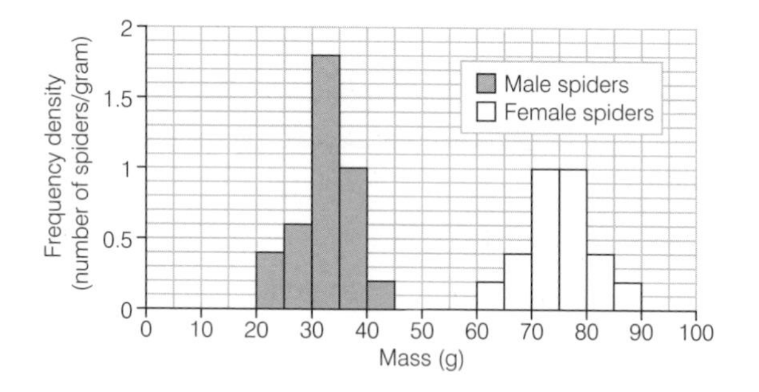

EXTENSION

Proof of coding

If we have data values $x_1, x_2, \ldots, x_n$ occuring with frequencies $f_1, f_2, \ldots, f_n$ and new data value $y_1, \ldots, y_n$ are obtained from the old using the linear rule

$$y = ax + b$$

then

$$\bar{y} = \frac{\sum yf}{N} = \frac{y_1 f_1 + y_2 f_2 + \cdots + y_n f_n}{N}$$

$$\Rightarrow \quad \bar{y} = \frac{(ax_1 + b)f_1 + (ax_2 + b)f_2 + \cdots + (ax_n + b)f_n}{N}$$

$$\Rightarrow \quad \bar{y} = \frac{a(x_1 f_1 + x_2 f_2 + \cdots + x_n f_n) + b(f_1 + f_2 + \cdots + f_n)}{N}$$

$$\Rightarrow \quad \bar{y} = a\frac{(x_1 f_1 + x_2 f_2 + \cdots + x_n f_n)}{N} + b\frac{(f_1 + f_2 + \cdots + f_n)}{N} = a\bar{x} + b\frac{N}{N} = a\bar{x} + b \qquad \text{as required.}$$

Now consider the variance:

$$\text{Variance of } y = \frac{\sum (y - \bar{y})^2 f}{N} = \frac{(y_1 - \bar{y})^2 f_1 + \cdots + (y_n - \bar{y})^2 f_n}{N}$$

$$\Rightarrow \quad \text{Variance of } y = \frac{((ax_1 + b) - (a\bar{x} + b))^2 f_1 + \cdots + ((ax_n + b) - (a\bar{x} + b))^2 f_n}{N}$$

$$\Rightarrow \quad \text{Variance of } y = \frac{(ax_1 - a\bar{x})^2 f_1 + \cdots + (ax_n - a\bar{x})^2 f_n}{N}$$

$$\Rightarrow \quad \text{Variance of } y = \frac{a^2(x_1 - \bar{x})^2 f_1 + \cdots + a^2(x_n - \bar{x})^2 f_n}{N} = \frac{a^2 \sum (x - \bar{x})^2 f}{N}$$

$$\Rightarrow \quad \text{Variance of } y = a^2 \times \text{Variance of } x$$

Finally,

$$\text{Standard deviation of } y = \sqrt{\text{Variance of } y} = \sqrt{a^2 \times \text{Variance of } x}$$

$$\Rightarrow \quad \text{Standard deviation of } y = a\sqrt{\text{Variance of } x}$$

$$\Rightarrow \quad \text{Standard deviation of } y = a \times \text{Standard deviation of } x$$

If a is negative, the results for mean and variance remain the same. However, the last three lines must be amended since $\sqrt{a^2} = -a$ when a is negative so

$$\text{standard deviation of } y = \sqrt{\text{variance of } y} = \sqrt{a^2 \times \text{variance of } x}$$

$$\Rightarrow \quad \text{standard deviation of } y = -a\sqrt{\text{variance of } x}$$

$$\Rightarrow \quad \text{standard deviation of } y = -a \times \text{standard deviation of } x$$

Having studied this chapter you should know

- that if new data values $y_1, y_2, y_3, \ldots, y_n$, are obtained from old data values $x_1, x_2, x_3, \ldots, x_n$ using a linear rule $y = ax + b$ $(a > 0)$ then

  $\bar{y} = a\bar{x} + b$
  Variance of $y = a^2 \times$ Variance of x
  Standard deviation of $y = a \times$ Standard deviation of x

- how to find the mean, variance and standard deviation of a set of data obtained by combining two sets of data whose mean and standard deviations are known.

REVISION EXERCISE

1 A class of 20 students take a test. The score, x, for each student was recorded by the teacher. The results are summarised by

$$\Sigma(x - 10) = 208 \; \Sigma(x - 10)^2 = 2716$$

i) Calculate the standard deviation of the 20 scores.
ii) Show that $\Sigma x^2 = 8876$.
iii) Two other students took the test later. Their scores were 18 and 16. Find the mean and standard deviation of all 22 scores.

(OCR Jan 2001 S1)

2 At a primary school 20 boys and 25 girls did a test. The boys' scores, b, are summarised by

$$\Sigma b = 466 \qquad \Sigma b^2 = 11\,834$$

The mean of the girls' scores is 21.88 and the standard deviation of the girls' scores is 7.929, correct to 3 decimal places.

i) Find the mean of the boys' scores.
ii) Find the standard deviation of the boys' scores.
iii) Find the mean score of all 45 students.
iv) Find the standard deviation of the scores of all 45 students.
v) A box and whisker plot of the boys' and girls' scores is shown below.

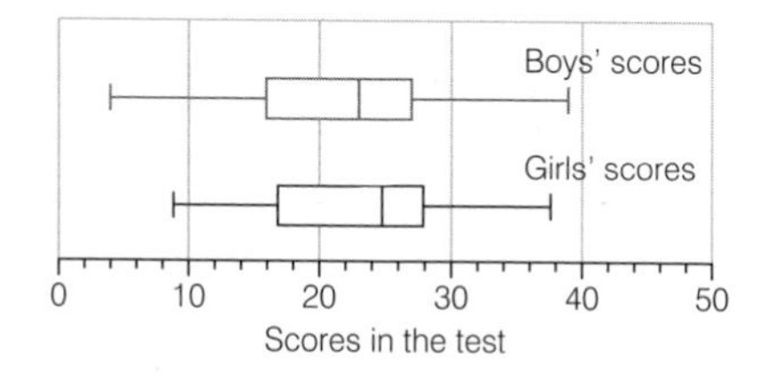

a) State the gender of the student who got the highest score in the test and give the corresponding score.
b) Use the box and whisker to make two statements comparing the overall performances of boys and girls in the test.

(OCR Nov 2002 S1)

3 A student carried out a statistics survey in a supermarket. At a checkout she recorded how many items each customer had in their shopping baskets. The table below gives her results.

Number of items in the basket	4	5	6	7	8	9
Number of customers with that number of items	2	7	9	11	2	5

i) Calculate the mean and standard deviation of the number of items the customers had in their shopping baskets.

ii) At the checkout each customer was given two free items. State the mean and standard deviation of the total number of items the customers now had.

(OCR Jan 2002 S1)

4 A "weights and measures" inspector has investigated the masses, m grams, of the contents of 12 jars of jam. His results can be summarised as

$$\Sigma(m - 450) = 78 \qquad \Sigma(m - 450)^2 = 555$$

a) Find the mean mass of the contents of the 12 jars.

b) Find the standard deviation of the masses of the contents of the 12 jars.

c) Calculate the value of $\Sigma(m + 175)$.

5 The histogram below shows the ages of the 66 cars parked in a factory's car park.

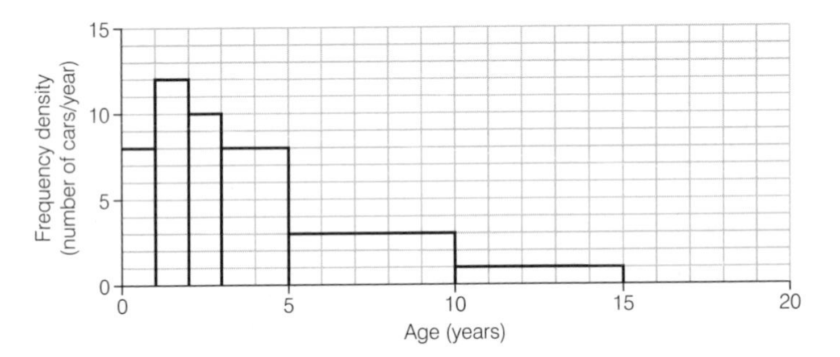

a) Produce a grouped frequency table for the data.

b) Calculate estimates of the mean and standard deviation of the data. Explain why your answers must be regarded as estimates.

The ages of 50 cars parked in a supermarket car park had mean 5.8 years and standard deviation 3.6 years.

c) Write a paragraph comparing the ages of the cars parked in the factory car park with the ages of the cars parked in the supermarket car park.

d) Calculate estimates of the mean and standard deviation of the ages of the combined sample of 116 cars.

6 The data collected in a survey are summarised by

$$\Sigma f = 80 \qquad \Sigma(5x - 4)f = 16 \qquad \Sigma(5x - 4)^2 f = 208$$

Obtain the mean and standard deviation of the data.

7 A school is planning a weekend trip to France and the pupils going on the exchange have been asked to indicate the amount of spending money they intend to take with them. The cumulative frequency diagram shows the results of the survey.

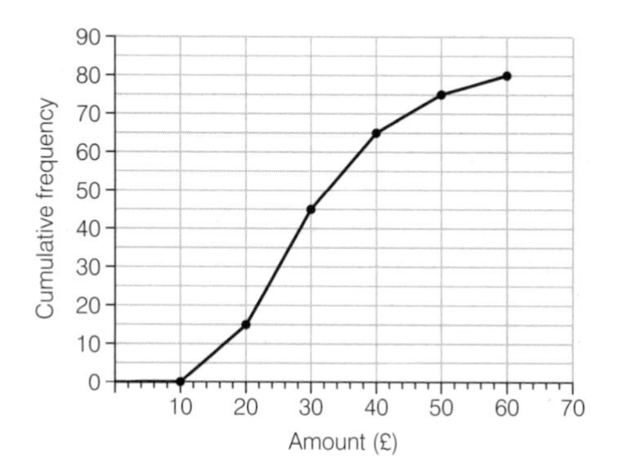

a) Obtain estimates of the median and interquartile range of the data.
b) Obtain estimates for the mean and standard deviation of the data.

The pupils will individually change their money at a bank that offers an exchange rate of £1 = €1.45 but deducts €3 as a service charge for the transaction.

A pupil who changes £x at this bank receives €y.

c) Obtain a relationship linking y to x.
d) Use your earlier answers to deduce estimates of:

i) the mean and standard deviation of the money in € that the pupils will have to spend in France;
ii) the median and interquartile range of the money in € that the pupils will have to spend in France.

8 A computer retailer has 40 desktop computers and 25 laptop computers for sale. The desktop computers have prices distributed with a mean of £650 and a standard deviation of £160, whilst the laptops have prices distributed with a mean of £845 and a standard deviation of £250. Find the mean and standard deviation of the prices of the 65 computers that the retailer has for sale.

4 Counting Methods and Probability

The purpose of this chapter is to enable you to

- use the multiplication, permutation and combination rules to calculate efficiently the number of ways complicated tasks may be performed
- use these rules in probability calculations

Counting Methods

You will encounter many situations where you wish to know the number of possible outcomes of an experiment. For the simplest experiments, such as tossing a coin or throwing a dice, it is easy to list the outcomes, but for most experiments this would be far too time-consuming. For example, trying to make a list of all the possible hands of five cards that can be dealt from a standard pack of cards would take a very long time. In the first section of this chapter three rules are developed that will enable you to calculate quickly the number of possible outcomes of some complex operations.

The Multiplication Rule

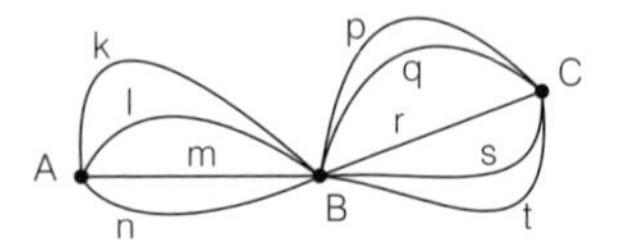

The diagram shows the roads linking towns A, B and C. There are four roads linking A to B, five linking B to C, but none linking A directly to C.

It should be clear that there are 20 routes from A to C via B:

k then p	l then p	m then p	n then p
k then q	l then q	m then q	n then q
k then r	l then r	m then r	n then r
k then s	l then s	m then s	n then s
k then t	l then t	m then t	n then t

This is a particular case of the **multiplication rule**, which states

If there are M different ways of doing operation 1
and N different ways of doing operation 2
then **there are $M \times N$ different ways of doing the two operations together**

EXAMPLE 1

A class contains 8 boys and 12 girls. The teacher wishes to choose one boy and one girl as class representatives. In how many different ways can he make the choice?

SOL

He can choose the boy in 8 different ways and he can choose the girl in 12 different ways, so he can choose the boy and the girl in $8 \times 12 = 96$ different ways.

The multiplication rule can be generalised to three (or more) operations:

If there are	M different ways of doing operation 1
	N different ways of doing operation 2
and	P different ways of doing operation 3
then **there are**	$\boldsymbol{M \times N \times P}$ **different ways of doing the three operations together**

Permutations

When selecting people to play in a football team, the manager may decide to select the goalkeeper first, the left back second and carry on until he finally selects the centre forward. In this case the order of selection is certainly important: a team that has David playing in goal and Michael playing as centre forward is certainly different from having Michael playing in goal and David playing as centre forward!

The number of **permutations** of r items from n is the number of ways of choosing r items from n when the order of choice **does** matter.

EXAMPLE 2

A classroom has 23 seats in it. In how many different ways can a class of 6 pupils sit in the classroom?

SOLUTION

Suppose the six pupils are A, B, C, D, E and F and this represents the order in which they enter the classroom.

A can sit in any one of 23 places	B can sit in any one of 22 places
C can sit in any one of 21 places	D can sit in any one of 20 places
E can sit in any one of 19 places	F can sit in any one of 18 places

So, by the multiplication rule, the class can sit in any one of

$23 \times 22 \times 21 \times 20 \times 19 \times 18 = 72\,681\,840$ different ways

The Factorial Function

The factorial function provides a useful shorthand for permutation problems.

If n is a positive integer then n factorial, written as $n!$, is defined by

$$n! = 1 \times 2 \times 3 \times \cdots \times n$$

so

$$1! = 1, \quad 2! = 1 \times 2 = 2, \quad 3! = 1 \times 2 \times 3 = 6, \quad 4! = 1 \times 2 \times 3 \times 4 = 24, \quad \text{etc.}$$

and

$0!$ is *defined* to be equal to 1.

The answer to Example 2 was obtained by calculating the value of $23 \times 22 \times 21 \times 20 \times 19 \times 18$.

This expression can be rewritten using factorial notation by noticing that

$$23 \times 22 \times 21 \times 20 \times 19 \times 18 = \frac{23 \times 22 \times 21 \times 20 \times 19 \times 18 \times 17 \times 16 \times \cdots \times 2 \times 1}{17 \times 16 \times \cdots \times 2 \times 1} = \frac{23!}{17!}$$

It has been found that the number of ways of choosing 6 seats from 23 when the order of choice is important is $23 \times 22 \times 21 \times 20 \times 19 \times 18$ or $\frac{23!}{17!}$.

As shorthand, we can say that ${}_{23}\text{P}_6$ is the number of ways of choosing 6 items from 23 when the order or choice is important, or that ${}_{23}\text{P}_6$ is the number of permutations of 6 items from 23. We can write

$${}_{23}\text{P}_6 = 23 \times 22 \times 21 \times 20 \times 19 \times 18 = \frac{23!}{17!}$$

In general:

The number of ways of choosing r items from n when the order of choice is important (**the number of permutations of r from n**) is

$${}_n\text{P}_r = n \times (n-1) \times (n-2) \times \cdots \times (n-r+1) = \frac{n!}{(n-r)!}$$

EXAMPLE 3

Peter has five tiles with numbers marked on them, as shown in the diagram.

3	4	7	8	9

He decides to choose three of the tiles and lay them out in a line to create a three-digit number. How many numbers can Peter make?

SOLUTION

Peter is choosing 3 tiles from 5 and the order of choice does matter (since 378 is different from 837).

Number of ways of doing this is ${}_5\text{P}_3 = \frac{5!}{(5-3)!} = \frac{5!}{2!} = 60$

Or
5 different choices for first tile;
4 different choices for second tile;
3 different choices for third tile.

So, number of possible choices for all three tiles:

$= 5 \times 4 \times 3 = 60$

by the multiplication rule.

Combinations

The permutation rule shows in how many ways choices of r objects from n can be made when the order of choice is important. The selection of a football team is an example of a choice where the order of choice is important. On the other hand, if a teacher wishes to select six pupils from a class to go on a theatre trip then the order of choice is unimportant.

The number of **combinations** of r items from n is the number of ways of choosing r items from n when the order of choice **does not** matter.

It is now possible to develop a rule for the number of ways of choosing r from n when the order of choice is unimportant.

EXAMPLE 4

Vicky has a bookcase with 23 books in it. She wishes to take 6 books on holiday with her. In how many different ways can she make her choice?

SOLUTION

You already know that there are ${}_{23}P_6 = 72\ 681\ 840$ different ways of choosing 6 books from 23 **if the order of choice is important**.

BUT in this example the order of choice is **NOT** important since, for example, the choice of books p, q, r, s, t, u is the same as the choice r, t, p, u, s, q.

In fact there are

$$6 \times 5 \times 4 \times 3 \times 2 \times 1 = 6!$$

different orderings of the 6 books p, q, r, s, t, u.

6 choices for the first book, 5 for the second, 4 for the third, 3 for the fourth, 2 for the fifth and 1 for the sixth.

So (multiplication rule) there are

$$6 \times 5 \times 4 \times 3 \times 2 \times 1 = 6!$$

possible orderings of the 6 books p, q, r, s, t, u.

Thus the 72 681 840 different ways of choosing 6 books from 23 if the order of choice is important repeats each selection of books 6! times, so the number of possible choices is

$$\frac{72\ 681\ 840}{6!} = 100\ 947$$

Notice that $\dfrac{72\ 681\ 840}{6!} = \dfrac{{}_{23}P_6}{6!} = \dfrac{23!/17!}{6!} = \dfrac{23!}{17! \times 6!}$

You have found that the number of ways of choosing 6 books from 23 when the order of choice is **not** important is $\frac{23!}{17! \times 6!}$.

As shorthand, ${}_{23}C_6$ or $\binom{23}{6}$ is written for the number of ways of choosing 6 items from 23 when the order of choice is not important. Alternatively, you can say that ${}_{23}C_6$ or $\binom{23}{6}$ is the number of combinations of 6 items from 23. You can write

$${}_{23}C_6 = \binom{23}{6} = \frac{23!}{17! \times 6!}$$

Generalising this argument, the following result is obtained:

The number of ways of choosing r items from n when the order of choice is not important (**the number of combinations of r from n**) is

$$\binom{n}{r} = {}_nC_r = \frac{n!}{(n-r)! \times r!}$$

The factorial function, permutations and combinations should be on your calculator. Make sure you know how to use them!

EXAMPLE 5

In how many different ways can a committee of 7 people be chosen from a group of 20 people?

SOL

The order of the choice is unimportant, so combinations can be used:

No. of different ways of choosing committee $= {}_{20}C_7 = 77\ 520$

EXAMPLE 6

Mr Jones teaches a class of 12 pupils.

a) In how many ways can he choose 5 pupils to stay behind after the lesson to tidy the classroom?

b) In how many ways can he choose 5 pupils and arrange them in a straight line?

SOLUTION

a) In this case the order of choice is unimportant, so combinations can be used:

No. of possible choices = ${}_{12}C_5 = 792$

b) In this case the order of choice does make a difference, since the line A, B, C, D, E is different from the line B, A, C, E, D. The permutation rule must be used:

No. of possible lines = ${}_{12}P_5 = 95\ 040$

EXERCISE 1

1 On a small island car number plates consist of a letter of the alphabet followed by a single-digit number chosen from 1, 2, 3, 4, 5, 6, 7, 8 and 9.
How many different possible number plates are there on this island?

2 In how many different ways can a boy and a girl be chosen from a group of 10 boys and 15 girls?

3 James has 5 different coloured shirts and 8 different ties. In how many different ways can he choose a shirt and tie to wear?

4 From the menu for a three-course meal at a restaurant a diner must choose one of 5 starters, one of 6 main courses and one of 4 desserts. How many different three-course meals is it possible to make from this menu?

5 A "string" consists of a sequence of letters. For example, PGYX is a four-letter string. Saffron has the 8 letter tiles shown below.

How many different four-letter strings can she make by choosing 4 different tiles?

6 Saffron has a set of 12 different coloured crayons. She has a picture showing a house, some ground and the sky, and she decides to colour the sky with one colour, the ground with a different colour and the house with a third colour. In how many different ways can she colour the picture?

7 In how many ways can a group of 5 people be chosen from a meeting of 24 people?

8 A country has 30 major landmarks. Mark is planning a visit to the country but knows he only has time to visit 8 of these landmarks. In how many different ways can he choose the 8 landmarks to visit?

9 A Sunday newspaper is split into 11 different sections. Michael decides to take 3 of the sections to read in the garden. In how many different ways can he choose these sections?

10 An investment company offers options to invest in 20 different unit trusts.

a) Michael decides he is going to invest £1000 in each of 3 of these unit trusts. In how many different ways can he invest his money?

b) Vicki decides she is going to invest £1500 in one unit trust, £1000 in a second unit trust and £500 in a third unit trust. In how many different ways can she invest her money?

Further Examples

Remember that
A standard pack of cards consists of 52 cards consisting of
Ace, 2, 3, 4, 5, 6, 7, 8, 9, 10, Jack, Queen and King in each of four suits:
Clubs, Diamonds, Hearts and Spades.

EXAMPLE 7

A hand of seven cards is dealt from a standard pack.

How many different possible hands are there?

How many of these hands consist of:

a) three clubs and four spades;
b) exactly five diamonds;
c) seven spades;
d) seven cards all in one suit?

SOLUTION

The order of the deal does not matter for most card games, so combinations can be used:

No. of different possible hands = No. of ways of choosing 7 cards from 52 when order doesn't matter

$= {}_{52}C_7 = 133\ 784\ 560$

a) There are ${}_{13}C_3$ different ways of obtaining 3 club cards from the 13 and ${}_{13}C_4$ different ways of obtaining 4 spade cards from the 13, so:

No. of hands containing 3 clubs and 4 spades $= {}_{13}C_3 \times {}_{13}C_4 = 204\ 490$

multiplication rule

b) There are ${}_{13}C_5$ different ways of obtaining 5 diamond cards from the 13 and ${}_{39}C_2$ different ways of obtaining 2 cards that are not diamonds from the 39 non-diamonds, so:

No. of hands containing exactly 5 diamonds $= {}_{13}C_5 \times {}_{39}C_2 = 953\ 667$

c) There are ${}_{13}C_7$ different hands containing 7 spades, so:

No. of hands containing 7 spades $= {}_{13}C_7 = 1716$

d) There are 1716 hands containing 7 spades, 1716 containing 7 hearts, 1716 containing 7 diamonds and 1716 containing 7 clubs, so:

No. of hands containing 7 cards of one suit $= 4 \times 1716 = 6864$

EXAMPLE 8

An Annual General Meeting of a society was attended by 16 men and 25 women. In how many ways can a committee of 5 be chosen from the people attending the meeting if:

i) there are no restrictions;
ii) the committee must have 2 men and 3 women;
iii) the committee must have at least 2 men and at least 2 women?

SOLUTION

The order of choice for the committee is not important, so combinations can be used.

i) No. of ways of choosing 5 from 41 = ${}_{41}C_5 = 749\,398$.
ii) No. of ways of choosing 2 men from 16 and 3 women from 25
$= {}_{16}C_2 \times {}_{25}C_3 = 276\,000$.
iii) To have at least 2 men and at least 2 women the committee must be 2 men, 3 women or 3 men, 2 women.
There are 276 000 ways of getting a committee of 2 men and 3 women and there are ${}_{16}C_3 \times {}_{25}C_2 = 168\,000$ ways of getting a committee of 2 men and 3 women.
Hence, there are 276 000 + 168 000 = 444 000 different ways of obtaining a committee with at least 2 men and at least 2 women.

EXERCISE 2

1 In how many different ways can 5 boys and 3 girls be chosen from a group of 8 boys and 6 girls?

2 **a)** At James's school, students must choose 4 AS subjects from a list of 18. In how many different ways can James make his choice?
b) At Anne's college, students must choose 1 subject from each of 4 different blocks of subjects. In Block A there are 4 different subjects; in Block B there are 6 different subjects; in Block C there are 5 different subjects and in Block D there are 4 different subjects. In how many different ways can Anne choose her 4 subjects?

3 **a)** Hayley has a set of 26 cards with each letter of the alphabet appearing on exactly 1 card. How many different hands of 5 cards can be dealt from this pack?
b) How many of these hands contain:
i) 5 consonants;
ii) exactly 3 consonants?

4 At a sportswomen's dinner there are 12 women who are hockey players and 15 other women who are netball players. In how many ways can 3 hockey players and 2 netball players be selected from the people attending the dinner?

5 A basket contains 25 balls. Each ball has an integer from 1 to 25 printed on it and no 2 balls have the same number printed on them. In how many different ways can 6 balls be taken from the basket:
a) with no restriction;
b) so that exactly 2 even numbered balls are taken;
c) no balls whose number is a multiple of 5 are taken?

6 A class of 25 pupils must each choose 1 subject from Art, Music and Drama. Ten of the pupils study Art, six study Music and nine study Drama. In how many different ways can four pupils be chosen from the class:

a) with no restriction;
b) in such a way that all four study Art;
c) in such a way that all four study the same subject;
d) in such a way that none study Drama;
e) in such a way that at least one studies Drama;
f) in such a way that exactly two study Art;
g) in such a way that two study Art, one studies Music and one studies Drama;
h) in such a way that there is at least one studying Art, at least one studying Music and at least one studying Drama?

7 **a)** How many different hands of 6 cards can be dealt from a standard pack of cards?
b) How many of these hands consist of:
i) 4 spades and 2 clubs;
ii) 4 spades and 2 cards of one other suit;
iii) 4 cards of one suit and 2 cards of another suit?

Basic Probability

Probability theory attempts to attach probabilities or likelihoods of events happening when a random experiment is performed. There are several key terms that must be understood from the outset.

A **random experiment** is an experiment whose outcome cannot be predicted in advance.

The **sample space** (or outcome space) of a random experiment is the set (or list) of all possible outcomes of the experiment.

An **event** is a subset of the sample space of the random experiment.

The table below illustrates these ideas for three simple random experiments.

	Random experiment	Sample space	Example of an event described verbally	Event described as a subset of the sample space
1	Toss a coin and record whether it lands on Heads or Tails.	{Heads, Tails}	Coin lands on Tails	{Tails}
2	Throw a cubical die with faces marked 1, 2, 3, 4, 5 and 6, and record the score on the uppermost face.	{1, 2, 3, 4, 5, 6}	The score is an even number	{2, 4, 6}
3	Pick a card at random from a standard pack and record its suit.	{Club, Diamond, Heart, Spade}	The card comes from a black suit	{Club, Spade}

A **probability model** is a way of measuring how likely an event is to occur. The measuring scale gives a probability of 0 to an impossible event and a probability of 1 to an event that is certain to occur.

A particularly important example of a probability model is the case when the sample space consists of outcomes that are equally likely to occur. In this case the rule we use to assign probabilities is:

$$P(\text{event}) = \frac{\text{number of ways in which the event can happen}}{\text{total number of outcomes of the experiment}}$$

EXAMPLE 9

A pair of fair cubical dice are thrown. What is the probability that the total score exceeds 9?

SOLUTION

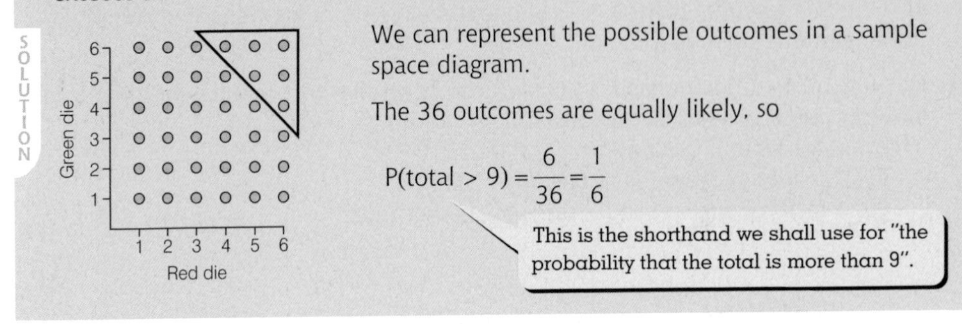

We can represent the possible outcomes in a sample space diagram.

The 36 outcomes are equally likely, so

$$P(\text{total} > 9) = \frac{6}{36} = \frac{1}{6}$$

This is the shorthand we shall use for "the probability that the total is more than 9".

The rule for combinations is very useful in the probability work that follows when dealing with experiments having a very large number of possible outcomes.

EXAMPLE 10

A hand of four cards is dealt from a standard pack of cards. Find the probability of the following events:

a) the hand contains four clubs;
b) the hand contains exactly two kings;
c) the hand contains exactly two kings and exactly one queen.

SOLUTION

Notice that the order in which the cards are dealt does **not** matter in this example.

There are ${}_{52}C_4 = 270\,725$ different hands of four cards.

a) There are ${}_{13}C_4 = 715$ different hands consisting of four clubs, so

$$P(4\text{ clubs}) = \frac{{}_{13}C_4}{{}_{52}C_4} = \frac{715}{270\,725} = 0.00264 \qquad (3\text{ s.f.})$$

b) If the hand is to contain two kings then it must contain two kings chosen from four AND two other cards chosen from the 48 cards in the pack that are NOT kings. This can be done in ${}_4C_2 \times {}_{48}C_2$ ways:

$$P(2\text{ kings}) = P(2\text{ kings and 2 others}) = \frac{{}_4C_2 \times {}_{48}C_2}{{}_{52}C_4} = \frac{6 \times 1128}{270\,725}$$

$$= \frac{6768}{270\,725} = 0.0250 \qquad (3\text{ s.f.})$$

EXAMPLE 10 (continued)

c) If the hand is to contain exactly two kings and exactly one queen then it must also contain one other card which is not a king or a queen. There are ${}_4C_2 \times {}_4C_1 \times {}_{44}C_1$ hands that meet this condition:

$$P(\text{2 kings and 1 queen}) = P(\text{2 kings and 1 queen and 1 other})$$
$$= \frac{{}_4C_2 \times {}_4C_1 \times {}_{44}C_1}{{}_{52}C_4} = \frac{6 \times 4 \times 44}{270\,725} = \frac{1056}{270\,725} = 0.00390 \quad \text{(3 s.f.)}$$

EXAMPLE 11

A class of 30 pupils contains 8 pupils who walk to school, 10 who cycle and 12 who are driven. The teacher chooses 4 pupils at random from the class. Find the probability that:

a) all four cycle to school;
b) all four come to school by the same means;
c) exactly two cycle;
d) two cycle, one walks and one is driven;
e) at least one cycles, at least one walks and at least one is driven.

SOLUTION

Again the order of choice does not matter, so combinations can be used.

There are ${}_{30}C_4$ ways of choosing four pupils from the class:

a) $P(\text{4 cycle}) = \frac{{}_{10}C_4}{{}_{30}C_4} = \frac{210}{27\,405} = 0.00766 \quad \text{(3 s.f.)}$

b) $P(\text{all 4 come by the same means}) = P(\text{4 cycle or 4 walk or 4 driven})$
$= P(\text{4 cycle}) + P(\text{4 walk}) + P(\text{4 driven})$

$$= \frac{{}_{10}C_4}{{}_{30}C_4} + \frac{{}_8C_4}{{}_{30}C_4} + \frac{{}_{12}C_4}{{}_{30}C_4} = 0.00766... + 0.00255... + 0.01806... = 0.0283 \quad \text{(3 s.f.)}$$

c) $P(\text{exactly 2 cycle}) = P(\text{2 cycle and 2 don't cycle})$

$$= \frac{{}_{10}C_2 \times {}_{20}C_2}{{}_{30}C_4} = 0.312 \quad \text{(3 s.f.)}$$

d) $P(\text{2 cycle, 1 walks, 1 driven}) = \frac{{}_{10}C_2 \times {}_8C_1 \times {}_{12}C_1}{{}_{30}C_4} = 0.158 \quad \text{(3 s.f.)}$

e) If you are to get at least one of each mode of transport there must be:

2 cycle, 1 walks, 1 driven
OR
1 cycles, 2 walk, 1 driven
OR
1 cycles, 1 walks, 2 driven

So

$$P(\text{at least one of each mode}) = P(\text{2C, 1W, 1D or 1C, 2W, 1D or 1C, 1W, 2D})$$
$$= \frac{{}_{10}C_2 \times {}_8C_1 \times {}_{12}C_1}{{}_{30}C_4} + \frac{{}_{10}C_1 \times {}_8C_2 \times {}_{12}C_1}{{}_{30}C_4} + \frac{{}_{10}C_1 \times {}_8C_1 \times {}_{12}C_2}{{}_{30}C_4}$$
$$= 0.1576... + 0.1226... + 0.1926... = 0.473 \quad \text{(3 s.f.)}$$

EXERCISE 3

1 A basket contains 5 red marbles, 3 yellow marbles and 2 green marbles. John takes a marble out of the basket at random. What is the probability that the marble he takes is:
a) a yellow marble; **b)** a red marble; **c)** not a green marble?

2 An octahedral die has 8 faces marked 1, 2, 3, 4, 5, 6, 7 and 8. When this die is thrown the score is the number on the face that lands pointing up. Ian throws an octahedral die. What is the probability that his score:
a) is 5, **b)** is an even number,
c) is a prime number, **d)** is not 8?

3 Two fair cubical dice, each with faces marked 1, 2, 3, 4, 5, 6, are thrown and the total score is recorded. Draw a suitable diagram to find the probability that the total score is:
a) 5, **b)** 9, **c)** an even number,
d) an odd number, **e)** a prime number, **f)** more than 8.

4 Richard has a fair cubical die with faces marked 1, 2, 3, 4, 5 and 6 and a fair octahedral die with faces marked 1, 1, 1, 2, 2, 3, 3 and 4. The two dice are thrown and the total score is calculated by adding the two scores together. Draw a diagram to show all the possible outcomes and find the probability that the total score is:
a) 8, **b)** an even number, **c)** at least 9.

5 A class contains 16 boys and 8 girls. Five pupils are to be chosen from the class at random to form a committee. Find the probability that:
a) the committee consists of 4 boys and 1 girl;
b) the committee consists of boys only;
c) the committee has a majority of boys.

6 A box of 12 biros contains 5 blue, 4 black and 3 red biros. John takes three biros, at random, out of the pack. Find the probability that John gets:
a) three blue biros;
b) three biros of the same colour;
c) exactly one black biro;
d) one biro of each colour.

7 A reduced pack of cards consists of

♣ A, K, Q, J ♦ A, K, Q
♥ A, K ♠ K, Q, J

A hand of three cards is dealt from this pack. Find the probability that the hand:
a) consists of three clubs;
b) consists of three cards from the same suit;
c) has exactly one ace;
d) has exactly one jack;
e) has exactly one ace and exactly one jack.

8 Four different numbers are to be picked at random from the set {1, 2, 3, 4, ..., 18, 19, 20}. Find the probability that:
a) the four numbers are all multiples of 3;
b) the four numbers are all less than 15;
c) the smallest number is 8;
d) the smallest number is 5 and the largest number is 16;
e) the product of the four numbers is an odd number.

Hint: this means that the 8 must be chosen and the other three numbers must be 9 or more.

9 At a party there are 18 people. 8 are Conservative supporters, 7 are Labour supporters and 3 are Liberal Democrats. In the corner of the room there is a group of four people. If this is a totally random group, find the probability that it contains:

a) no Liberals;
b) at least one Liberal;
c) exactly two Labour supporters;
d) two Conservatives, one Labour and one Liberal;
e) at least one supporter of each party.

Having studied this chapter you should know how to

- use the **multiplication rule:**

  If there are M different ways of doing operation 1
  and N different ways of doing operation 2
  then **there are $M \times N$ different ways of doing the two operations together**

- use the **permutation rule:**

  The number of ways of choosing r items from n when the order of choice is important (**the number of permutations of r from n**) is

  $$_n\mathrm{P}_r = n \times (n-1) \times (n-2) \times \cdots \times (n-r+1) = \frac{n!}{(n-r)!}$$

- use the **combination rule:**

  The number of ways of choosing r items from n when the order of choice is not important (**the number of combinations of r from n**) is

  $$\binom{n}{r} = {_n\mathrm{C}_r} = \frac{n!}{(n-r)! \times r!}$$

- use the basic language of probability. In particular you should know that:

  A **random experiment** is an experiment whose outcome cannot be predicted in advance.

  The **sample space** (or outcome space) of a random experiment is the set (or list) of all possible outcomes of the experiment.

  An **event** is simply a subset of the sample space of the random experiment.

  A **probability model** is a way of measuring how likely an event is to occur. The measuring scale gives a probability of 0 to an impossible event and a probability of 1 to an event that is certain to occur.

- use the formula

  $$\mathrm{P(event)} = \frac{\text{number of ways in which the event can happen}}{\text{total number of outcomes of the experiment}}$$

  to calculate probabilities of events when the sample space consists of equally likely outcomes.

REVISION EXERCISE

1 Seven men and five women have been nominated to serve on a committee. The committee consists of four members who are to be chosen from the seven men and five women.

i) In how many different ways can the committee be chosen?

ii) In how many of these ways will the committee consist of two men and two women?

iii) Assuming that each choice of four members is equally likely, find the probability that the committee will contain exactly two men.

(OCR Jan 2002 S1)

2 A bag contains 40 plastic tiles which are used in a word game. Each tile has a single letter written on it. 16 of the tiles have vowels written on them and the remaining 24 tiles have consonants written on them. Tara picks five tiles, at random and without replacement, out of the bag.

Find the probability that Tara chooses two tiles with vowels and three with consonants.

3 Brenda has a fair cubical die with faces marked 1, 2, 3, 4, 5 and 6 and a fair octahedral die with faces marked 1, 1, 1, 1, 2, 2, 3 and 4. The two dice are thrown and the total score is calculated by multiplying the two scores together.

Draw a diagram to show all the possible outcomes and find the probability that the total score is:

a) 6, **b)** an odd number, **c)** more than 10.

4 How many different hands of four cards can be dealt from a standard pack of cards?

How many of these different hands contain:

a) two spades, one heart and one diamond;

b) exactly two spades;

c) no spades;

d) at least one spade;

e) one card from each suit?

5 In how many different ways can four pupils be chosen from a class of 12 pupils if:

a) one pupil is to be sent to the theatre, one to the library, one to the church and one to the supermarket;

b) all four pupils are to go to the cinema together?

6 Vicky has a bag with 25 plastic tiles which are used in a number game. Each tile has a single whole number written on it and the numbers 1, 2, 3, ..., 24, 25 each appear on exactly one tile. If Vicky takes six tiles, at random and without replacement, out of the bag find the probability that she obtains:

a) four odd and two even numbers;

b) six numbers all of which are less than 19;

c) six numbers that multiply together to give an even number.

7 A box of fuses contains 8 thirteen amp fuses and 4 five amp fuses. Pauline picks three fuses at random out of the box. Find the probability that she picks:

a) 3 thirteen amp fuses;

b) 2 five amp fuses and 1 thirteen amp fuse.

8 A family consists of two parents and eight children, of whom three are boys. The family have been invited to send six representatives to a party. In how many ways can this be done if:

a) there are no restrictions;

b) exactly one adult and five children must go to the party;

c) one adult, three female children and two male children must go to the party?

9 A committee of 6 people is to be elected from a group of 8 men and 6 women.

a) How many possible committees are there?

b) How many different committees have two men and four women?

c) Prove that there are 2590 different committees that have at least two male and at least two female members.

d) If the committee is chosen at random find the probability that it has at least two male and at least two female members.

10 a) In how many different ways can a group of 20 girls be divided into a group of 8 and a group of 12?

b) In how many different ways can a group of 20 girls be divided into a group of 8, a group of 7 and a group of 5?

5 Laws of Probability

The purpose of this chapter is to enable you to

- use the basic laws of probability
- understand and use conditional probability
- work with independent events

The previous chapter showed how the probabilities of relatively simple events can be calculated. You now know, for example, how to calculate the probability that 2 men and 3 women are chosen if 5 people are selected at random from a group of 10 men and 12 women. However, as yet you have no simple means of calculating the probability of obtaining at least 2 women when this selection is made. In this chapter general rules of probability will be introduced and these rules will dramatically increase the range of problems that can be answered.

P(A) + P(not A)

The probability of throwing a six with a fair cubical die is $\frac{1}{6}$; the probability of not throwing a six with a fair cubical die is $\frac{5}{6}$.

If Saffron has probability 0.21 of catching a cold during the next month then she has probability 0.79 of not catching a cold during the next month.

In general, if A is any event then the probability that A happens added to the probability that A does not happen must give 1.

If we write "not A" as shorthand for "Event A doesn't happen" then we have the rule

P(A) + P(not A) = 1

This can be rewritten as

Rule 1
P(A) = 1 − P(not A)

EXAMPLE 1

John has a bag with 5 blue biros, 4 black biros and 3 red biros. If he takes 3 biros out of the bag at random find the probability that he gets at least 1 blue biro.

EXAMPLE 1 (continued)

SOLUTION

If

A is the event "John gets at least 1 blue biro"

then

Not A is the event "John does not get at least 1 blue biro". This is the same as saying "John gets no blue biros".

$$P(\text{at least 1 blue}) = 1 - P(\text{no blues})$$

$$= 1 - \frac{{}_7C_3}{{}_{12}C_3} = 1 - 0.1591 = 0.841 \quad (3 \text{ s.f.})$$

There are ${}_{12}C_3$ different ways of choosing 3 biros from the 12. There are 7 biros that are not blue, so there are ${}_7C_3$ different ways of choosing 3 non-blue biros.

Hence $P(\text{no blue biros}) = \frac{{}_7C_3}{{}_{12}C_3}$

P(A or B)

Interpreting the Word "or"

In common usage the word "or" can be ambiguous.

The statement "9 is an odd number or a prime number" is true because 9 is odd.
The statement "2 is an odd number or a prime number" is true because 2 is prime.
The statement "12 is an odd number or a prime number" is clearly false.

The ambiguity is clear when you consider the statement

"13 is an odd number or a prime number."

Some people will argue that this statement is false on the basis that "or" means **just one of the conditions is satisfied**.

Others will argue that this statement is true on the basis that "or" means **at least one of the conditions is satisfied**.

In Mathematics the convention is that "or" means **at least one of the conditions is satisfied**.

So, in Mathematics, all the following statements are true:

"13 is an odd number or a prime number."
"100 is a square number or an even number."
"The letter R appears in the word JANUARY or in the word MARCH."

The Rules for P(A or B)

Consider the experiment of throwing two fair dice – one red and one green.

Let A be the event "the total score is 4" and B be the event "the score on the red die is 6".

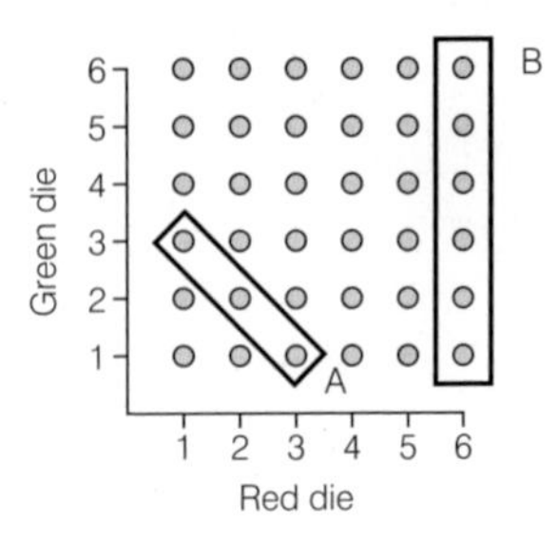

From the diagram you can see that

$$P(A) = \frac{3}{36}, \qquad P(B) = \frac{6}{36}$$

and

$$P(A \text{ or } B) = \frac{9}{36} = P(A) + P(B)$$

Now let A be the event "the total score is 4" and let C be the event "the score on the red die is 3".

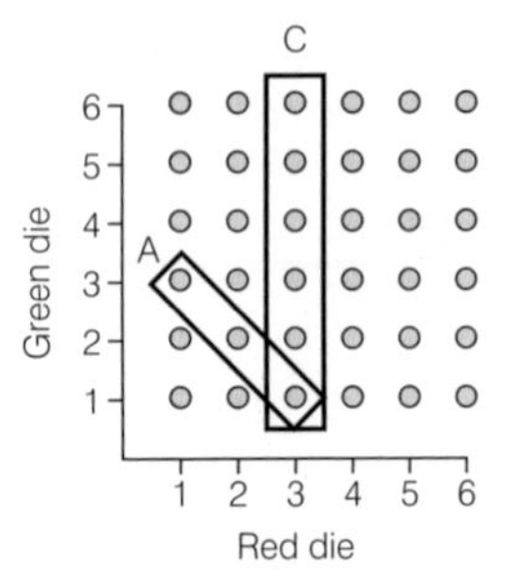

From the diagram you can see that

$$P(A) = \frac{3}{36}, \qquad P(C) = \frac{6}{36}$$

and

$$P(A \text{ or } C) = \frac{8}{36} \neq P(A) + P(C)$$

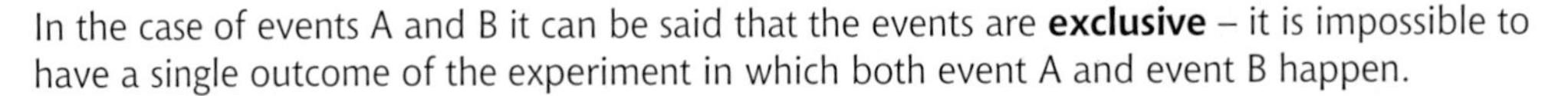

In the case of events A and B it can be said that the events are **exclusive** – it is impossible to have a single outcome of the experiment in which both event A and event B happen.

> **Rule 2**
> If A and B are exclusive then
>
> **P(A or B) = P(A) + P(B)**

In the case of sets A and C the events are not exclusive – it is possible to have an outcome of the experiment in which both the event A and the event C happen (red = 3, green = 1).

In this case you will see that

$$P(A \text{ or } C) = \frac{8}{36} = \frac{3}{36} + \frac{6}{36} - \frac{1}{36} = P(A) + P(C) - P(A \text{ and } C)$$

> **Rule 3**
> If A and B are not exclusive then
>
> **P(A or B) = P(A) + P(B) – P(A and B)**

EXAMPLE 2

A hand of 6 cards is dealt from a standard pack of cards. Find the probability that the hand has:

a) 5 or 6 clubs;
b) less than 5 clubs.

EXAMPLE 2 (continued)

SOLUTION

a) $P(\text{5 clubs}) = P(\text{5 clubs and 1 non-club}) = \dfrac{{}_{13}C_5 \times {}_{39}C_1}{{}_{52}C_6} = 0.002465\ldots$

$P(\text{6 clubs}) = \dfrac{{}_{13}C_6}{{}_{52}C_6} = 0.000084\ldots$

$P(\text{5 or 6 clubs}) = P(\text{5 clubs}) + P(\text{6 clubs}) = 0.002465\ldots + 0.000084\ldots = 0.00255$ (3 s.f.)

b) $P(\text{less than 5 clubs}) = P(0, 1, 2, 3 \text{ or } 4 \text{ clubs})$
$= 1 - P(\text{5 or 6 clubs})$
$= 1 - 0.00255$
$= 0.997$ (3 s.f.)

Rule 1

Use **rule 2** since the events "5 clubs" and "6 clubs" are exclusive. No single hand of 6 cards can possibly contain exactly 5 club cards and exactly 6 club cards.

EXAMPLE 3

A hand of 4 cards is dealt from a standard pack. Find the probability of obtaining:

a) one or two diamonds;
b) exactly two kings;
c) exactly one ace;
d) exactly two kings and exactly one ace;
e) exactly two kings or exactly one ace.

SOLUTION

a) The events "exactly 1 diamond" and "exactly 2 diamonds" are exclusive, so

$P(\text{1 or 2 diamonds}) = P(\text{1 diamond}) + P(\text{2 diamonds})$
$= P(\text{1 diamond, 3 others}) + P(\text{2 diamonds, 2 others})$
$= \dfrac{{}_{13}C_1 \times {}_{39}C_3}{{}_{52}C_4} + \dfrac{{}_{13}C_2 \times {}_{39}C_2}{{}_{52}C_4} = 0.6523\ldots = 0.652$ (3 s.f.)

Rule 2

b) $P(\text{2 kings}) = P(\text{2 kings, 2 others}) = \dfrac{{}_{4}C_2 \times {}_{48}C_2}{{}_{52}C_4} = 0.02499\ldots = 0.0250$ (3 s.f.)

c) $P(\text{1 ace}) = P(\text{1 ace, 3 others}) = \dfrac{{}_{4}C_1 \times {}_{48}C_3}{{}_{52}C_4} = 0.25555\ldots = 0.256$ (3 s.f.)

d) $P(\text{2 kings and 1 ace}) = P(\text{2 kings, 1 ace, 1 other})$
$= \dfrac{{}_{4}C_2 \times {}_{4}C_1 \times {}_{44}C_1}{{}_{52}C_4} = 0.003900\ldots = 0.00390$ (3 s.f.)

NOTE this is **not** the same as

$P(\text{2 kings}) \times P(\text{1 ace})$

You need to be **very careful** when using the "rule"

$P(A \text{ and } B) = P(A) \times P(B)$

that you may have met in earlier studies.

The rules for P(A and B) will be considered later in the chapter. Until you have finished Exercise 3, you should **not** at any stage need to multiply two probabilities.

EXAMPLE 3 (continued)

e) The events "2 kings" and "1 ace" are not exclusive – it is quite possible to have a hand that contains 2 kings and also contains 1 ace. Therefore:

P(2 kings or 1 ace) = P(2 kings) + P(1 ace) – P(2 kings and 1 ace)
$= 0.02499\ldots + 0.25555\ldots - 0.00390\ldots$
$= 0.277$ (3 s.f.)

Rule 3

These probabilities have already been calculated in parts (b), (c) and (d).

EXAMPLE 4

At a large sixth form college, 45% of the pupils study Mathematics, 40% study Chemistry and 58% study either Mathematics or Chemistry.

If a student is picked at random from the college, find the probability that he studies:

a) Mathematics and Chemistry;
b) Chemistry but not Mathematics.

SOLUTION

a) The events "the student studies Maths" and "the student studies Chemistry" are not exclusive since a single student can easily study both subjects.

P(he studies M or C) = P(he studies M) + P(he studies C) – P(he studies M and C)
$\Rightarrow$ 0.58 = 0.45 + 0.40 – P(he studies M and C)
$\Rightarrow$ P(he studies M and C) = 0.45 + 0.40 – 0.58
$\Rightarrow$ P(he studies M and C) = 0.27

Rule 3

b) The relationship between the different subjects in a diagram can now be shown:

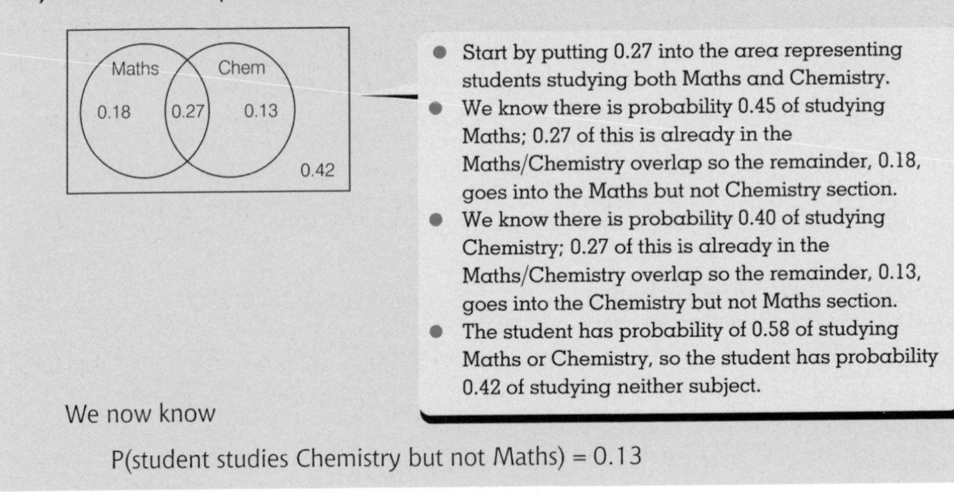

We now know

P(student studies Chemistry but not Maths) = 0.13

EXERCISE 1

1 An opinion poll claims that 27% of the population are Conservative supporters; 43% of the population are Labour supporters; 15% of the population are Liberal Democrat supporters.
Assuming these figures are correct, if I pick a person at random what is the probability that:

a) she will **not** be a Conservative supporter;
b) she will be a Conservative or a Labour supporter?

2 A market research survey claims that 32% of cars on the roads are British made; 23% of the cars on the roads are made in Japan; 28% of the cars on the roads are coloured red. Assuming these figures are correct, find the probability that the next car you see:

a) will not be red;

b) will have been made in Britain or Japan.

3 A crate of oranges contains 40 good oranges and 10 damaged oranges. The greengrocer gets 5 of the oranges out of the box at random to sell to Mrs Jones. Find the probability that:

a) exactly two are damaged;

b) at least one is damaged.

4 Every pupil in a class of 24 studies exactly one subject chosen from Greek or Music or Russian. Five of the pupils do Music and 11 do Russian. Four pupils are chosen at random from the class. Find the probability that:

a) exactly two do Russian;

b) at least one does Music;

c) at most three do Greek;

d) two do Greek, one does Russian and one does Music;

e) there is at least one pupil studying each subject in the group.

5 A pack of 16 plain cards is coloured in by John. Four are coloured red, 8 are coloured green and 4 are coloured blue. John shuffles the cards and deals a hand of three cards to Jane. Find the probability that:

a) the hand contains exactly 2 green cards;

b) the hand contains exactly 1 red card;

c) the hand contains 1 red card and 2 green cards;

d) the hand contains 1 red card or 2 green cards.

6 On a stretch of road there is a set of traffic lights and a pedestrian crossing quite close to each other. Experience suggests that the probability of being stopped at the traffic lights is 0.6; the probability of being stopped at the pedestrian crossing is 0.5; the probability of being stopped at the traffic lights or at the pedestrian crossing is 0.85. Find the probability of:

a) being stopped at the traffic lights and the pedestrian crossing;

b) being stopped at the pedestrian crossing but not at the traffic lights.

Conditional Probability

Suppose that John has a small pack of cards consisting of the ace, 2, 3, 4 and 5 of clubs, the ace and 2 of diamonds, the ace, 2 and 3 of hearts, and the 3 and 4 of spades:

John picks a card at random from the pack and asks you what the probability is that the card is a club.

Since there are 12 cards, 5 of which are clubs, the answer is $\frac{5}{12}$.

John now tells you that the card he has chosen is an ace and then asks you what the probability is that the card is a club. This time the answer is $\frac{1}{3}$, since there are three aces, one of which is also a club.

The second example is a case of **conditional probability** – some added information has been given about the outcome of the experiment and this affects our probability calculations.

You can write

$$\text{P(club} \mid \text{ace)} = \tfrac{1}{3}$$

The | stands for "given that".

as shorthand for the statement

"the probability that the card is a club **given that** it is an ace is $\frac{1}{3}$"

which can also be expressed as

"**if** the card is an ace **then** there is a probability of $\frac{1}{3}$ of it being a club"

Working with the same pack of cards:

$$\text{P(card is a 3)} = \tfrac{3}{12}, \qquad \text{P(card is a spade)} = \tfrac{2}{12}$$

and

$$\text{P(card is a 3} \mid \text{card is a spade)} = \tfrac{1}{2}$$

Notice that

$$\text{P(card is a 3 and card is a spade)} = \text{P(card is 3 of spades)} = \tfrac{1}{12}$$

and that

$$\text{P(card is a 3} \mid \text{card is a spade)} = \tfrac{1}{2} = \frac{\frac{1}{12}}{\frac{2}{12}} = \frac{\text{P(card is a 3 and card is a spade)}}{\text{P(card is a spade)}}$$

EXERCISE 2

This exercise considers some more examples of conditional probabilities and suggests the general rule that will be introduced in the next section.

You need to work through this exercise before reading the next section.

In questions 1–3, continue to consider the experiment of choosing one card from the small pack of cards consisting of

♣ : ace, 2, 3, 4, 5 ♦ : ace, 2

♥ : ace, 2, 3 ♠ : 3, 4

1 **a)** Find the values of:

i) P(card comes from a red suit);
ii) P(card is an ace);
iii) P(card comes from a red suit and card is an ace);
iv) P(card comes from a red suit | card is an ace).

b) Is it true that:

$$\text{P(card comes from a red suit} \mid \text{card is an ace)} = \frac{\text{P(card comes from a red suit and card is an ace)}}{\text{P(card is an ace)}}?$$

2 **a)** Find the values of:

i) P(card is a 4);
ii) P(card comes from a black suit);
iii) P(card is a 4 and card comes from a black suit);
iv) P(card is a 4 | card comes from a black suit).

b) Is it true that:

$$\text{P(card is a 4} \mid \text{card comes from a black suit)} = \frac{\text{P(card is a 4 and card comes from a black suit)}}{\text{P(card comes from a black suit)}}?$$

3 **a)** Find the values of:

i) P(card is a 4);
ii) P(card comes from a red suit);
iii) P(card is a 4 and card comes from a red suit);
iv) P(card is a 4 | card comes from a red suit).

b) Is it true that:

$$\text{P(card is a 4} \mid \text{card comes from a red suit)} = \frac{\text{P(card is a 4 and card comes from a red suit)}}{\text{P(card comes from a red suit)}}?$$

4 Hayley has two fair cubical dice, each with faces marked 1, 2, 3, 4, 5 and 6. One is coloured red and one is coloured blue. She throws the two dice. Draw a sample space diagram to show all the possible outcomes.

a) Find the values of:

i) P(total score is more than 9);
ii) P(score on red die is 5);
iii) P(total score is more than 9 and score on red die is 5);
iv) P(total score is more than 9 | score on red die is 5).

b) Is it true that:

$$\text{P(total score is more than 9} \mid \text{score on red die is 5)} = \frac{\text{P(total score is more than 9 and score on red die is 5)}}{\text{P(score on red die is 5)}}?$$

5 Hayley has two fair cubical dice, each with faces marked 1, 2, 3, 4, 5 and 6. One is coloured red and one is coloured blue. She throws the two dice. Draw a sample space diagram to show all the possible outcomes.

a) Find the values of:

i) P(total score is 4);
ii) P(score on red die is 5);
iii) P(total score is 4 and score on red die is 5);
iv) P(total score is 4 | score on red die is 5).

b) Is it true that:

$$\text{P(total score is 4} \mid \text{score on red die is 5)} = \frac{\text{P(total score is 4 and score on red die is 5)}}{\text{P(score on red die is 5)}}?$$

Working with Conditional Probabilities

You have seen that, for simple examples, you can use common sense to write down the values of conditional probabilities, and the answers to Exercise 2 suggest that there is a general rule for conditional probabilities:

Rule 4

$$P(A|B) = \frac{P(A \text{ and } B)}{P(B)}$$

This section will show how both approaches can be used to calculate some conditional probabilities.

EXAMPLE 5

A hand of six cards is dealt from a standard pack. Given that the hand contains exactly two aces, what is the probability that it contains the ace of spades?

SOLUTION 1

Common sense approach

The hand contains two aces, so the possibilities are:

ace of clubs, ace of diamonds
ace of clubs, ace of spades
ace of diamonds, ace of spades
ace of clubs, ace of hearts
ace of diamonds, ace of hearts
ace of hearts, ace of spades

Each of these possibilities are equally likely, so it is possible to write

$$P(\text{ace of spades} \mid \text{exactly two aces}) = \tfrac{3}{6} = \tfrac{1}{2}$$

SOLUTION 2

Alternatively, using rule 4

$$P(\text{ace of spades} \mid \text{exactly two aces}) = \frac{P(\text{ace of spades and exactly two aces})}{P(\text{exactly two aces})}$$

$$\Rightarrow \quad P(\text{ace of spades} \mid \text{exactly two aces}) = \frac{0.028673\ldots}{0.057346\ldots} = 0.5$$

since

$$P(\text{ace of spades and exactly two aces}) = P(\text{ace of spades, one other ace, four other cards})$$

$$= \frac{{}_1C_1 \times {}_3C_1 \times {}_{48}C_4}{{}_{52}C_6} = 0.028673\ldots$$

and

$$P(\text{exactly two aces}) = P(\text{two aces and four others}) = \frac{{}_4C_2 \times {}_{48}C_4}{{}_{52}C_6} = 0.057346\ldots$$

Clearly, in this example, it was much easier to use the "common sense approach", but this will not always be possible.

EXAMPLE 6

A bag contains 4 red, 7 yellow and 9 green marbles. Three of the marbles are taken at random out of the bag. Given that the 3 marbles are all the same colour, what is the probability that they are all yellow?

SOLUTION

Why would it be wrong to say that there are three possibilities RRR or YYY or GGG and therefore the desired probability is 1/3?

Rule 4

$$\text{P(3 yellow | all 3 are same colour)} = \frac{\text{P(3 yellow and all 3 are same colour)}}{\text{P(all 3 are same colour)}}$$

$$\Rightarrow \quad \text{P(3 yellow | all 3 are same colour)} = \frac{\text{P(3 yellow)}}{\text{P(all 3 are same colour)}}$$

If it is known that the 3 marbles are yellow **and** that the 3 marbles are the same colour, then this means the 3 marbles are yellow.

$$\Rightarrow \quad \text{P(3 yellow | all 3 are same colour)} = \frac{0.030701\ldots}{0.107894\ldots}$$

$$= 0.285 \quad \text{(3 s.f.)}$$

since

$$\text{P(3 yellow)} = \frac{{}_7C_3}{{}_{20}C_3} = 0.030701\ldots$$

and

$$\text{P(all 3 are same colour)} = \text{P(3 red)} + \text{P(3 yellow)} + \text{P(3 green)}$$

$$= \frac{{}_4C_3}{{}_{20}C_3} + \frac{{}_7C_3}{{}_{20}C_3} + \frac{{}_9C_3}{{}_{20}C_3} = 0.107894\ldots$$

EXERCISE 3

1 A group of 5 people is to be chosen, at random, from a group containing 4 Labour supporters, 6 Conservative supporters and 3 Liberal supporters. Find:
- **a)** the probability that the group contains exactly 3 Conservatives;
- **b)** the probability that the group contains 3 Conservatives, 1 Labour and 1 Liberal;
- **c)** the probability that the group contains exactly 1 Liberal given that the group contains exactly 3 Conservatives.

2 Paul has a box of 7 blue and 3 red biros. He takes 3 biros at random out of the box. Find:
- **a)** the probability of getting exactly 2 blue biros;
- **b)** the probability of getting at least 1 biro of each colour;
- **c)** the probability that he has 2 blue biros if it is known that he has at least 1 biro of each colour.

3 A hand of 3 cards is dealt from a standard pack of cards. Find the probability that:
- **a)** the hand contains at least 1 club;
- **b)** the hand contains 2 spades and 1 club;
- **c)** the hand contains 2 spades if it is known that the hand contains at least 1 club.

4 Saffron has 12 red bricks and 8 green bricks in a toy box. She picks 3 of the bricks at random out of the box.

a) Prove the probability that the 3 bricks chosen by Saffron are the same colour is $\frac{23}{95}$.

b) Find the probability that the 3 bricks chosen by Saffron are all red given that they are all the same colour.

5 A committee of 5 is to be chosen at random from a group of 8 men and 10 women.

a) Prove that the probability of obtaining a committee with at least 2 men and at least 2 women is $\frac{35}{51}$.

b) Find the probability that the committee contains 3 men and 2 women given that it has at least 2 men and at least 2 women.

The Multiplication Rule for Probabilities

You have seen that the rule for conditional probabilities is

$$P(D \mid C) = \frac{P(D \text{ and } C)}{P(C)}$$

and this formula can easily be rearranged to give

Rule 5
$P(C \text{ and } D) = P(C) \times P(D \mid C)$

This is known as the MULTIPLICATION RULE for probabilities.

You have probably already been using the multiplication rule in your earlier studies. Consider, for example, the following problem.

EXAMPLE 7

A bag contains 6 red and 3 blue marbles. Two marbles are taken at random. What is the probability that they are both red?

SOLUTION 1

Three solutions will be given to the problem. The first may be familiar from earlier work on probability – a tree diagram is used to show all the possible outcomes of the experiment in a systematic way and then the probability is calculated using the multiplication rule.

Let R_1 be shorthand for the first marble is red, with similar meanings for R_2, B_1 and B_2.

$\frac{6}{9}$ R_1: $\frac{5}{8}$ R_2 → $P(R_1 R_2) = \frac{6}{9} \times \frac{5}{8} = \frac{5}{12}$; $\frac{3}{8}$ B_2

$\frac{3}{9}$ B_1: $\frac{6}{8}$ R_2; $\frac{2}{8}$ B_2

Strictly speaking, the 5/8 is the CONDITIONAL PROBABILITY of the second marble being red given that the first marble is red.

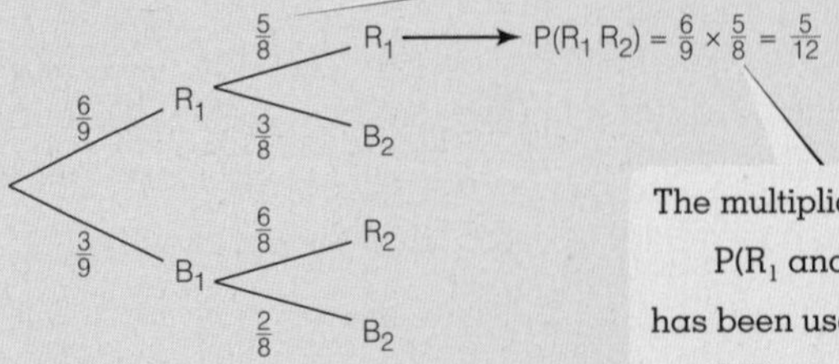

The multiplication rule:

$$P(R_1 \text{ and } R_2) = P(R_1) \times P(R_2 \mid R_1)$$

has been used to write down

$$P(R_1 \text{ and } R_2) = \tfrac{6}{9} \times \tfrac{5}{8} = \tfrac{5}{12}$$

EXAMPLE 7 (continued)

SOLUTION 2

At this level, the tree diagram may well not be necessary so a much shorter solution can be produced.

$$P(\text{both red}) = P(R_1 \text{ and } R_2) = \tfrac{6}{9} \times \tfrac{5}{8} = \tfrac{5}{12}$$

Just remember that

$\tfrac{6}{9} = P(R_1)$ and $\tfrac{5}{8} = P(R_2 | R_1)$

and that you are using the multiplication rule.

SOLUTION 3

It is, of course, possible to use combinations to answer this question.

$$P(\text{two red}) = \frac{{}_6C_2}{{}_9C_2} = \frac{15}{36} = \frac{5}{12}$$

EXAMPLE 8

In a multiple choice test each question has five possible answers. John has revised 60% of the subject and if there is a question on this part of the subject then he has a 90% chance of getting it correct. He guesses the answers to other questions.

a) Find the probability that John gets question 1 of the test correct.
b) If John got question 1 correct what is the probability that he had revised this subject?

SOLUTION 1

a) P(gets answer correct) = P(question is on subject he revised AND he gets it correct OR he has to guess AND guesses correctly)

$$= 0.6 \times 0.9 + 0.4 \times 0.2$$
$$= 0.54 + 0.08$$
$$= 0.62$$

The question tells us that:

P(question is on subject he revised) = 0.6
P(he has to guess) = 1 − 0.6 = 0.4
P(answers correctly | question is on topic revised) = 0.9
P(answers correctly | he has to guess) = 0.2

Multiplication rule gives:

P(knows subject and gets correct)
= P(knows subject) × P(gets correct | knows subject)
= 0.6 × 0.9 = 0.54

b) You need to find the conditional probability that he knew the subject of question 1 given that he got question 1 correct.

$$P(\text{revised subject} | \text{got question correct}) = \frac{P(\text{revised subject and got question correct})}{P(\text{got question correct})}$$

Rule 4

$$= \frac{0.6 \times 0.9}{0.62} = 0.871 \qquad (3 \text{ s.f.})$$

Again, this example could be answered with the aid of a tree diagram.

EXAMPLE 8 (continued)

SOLUTION 2

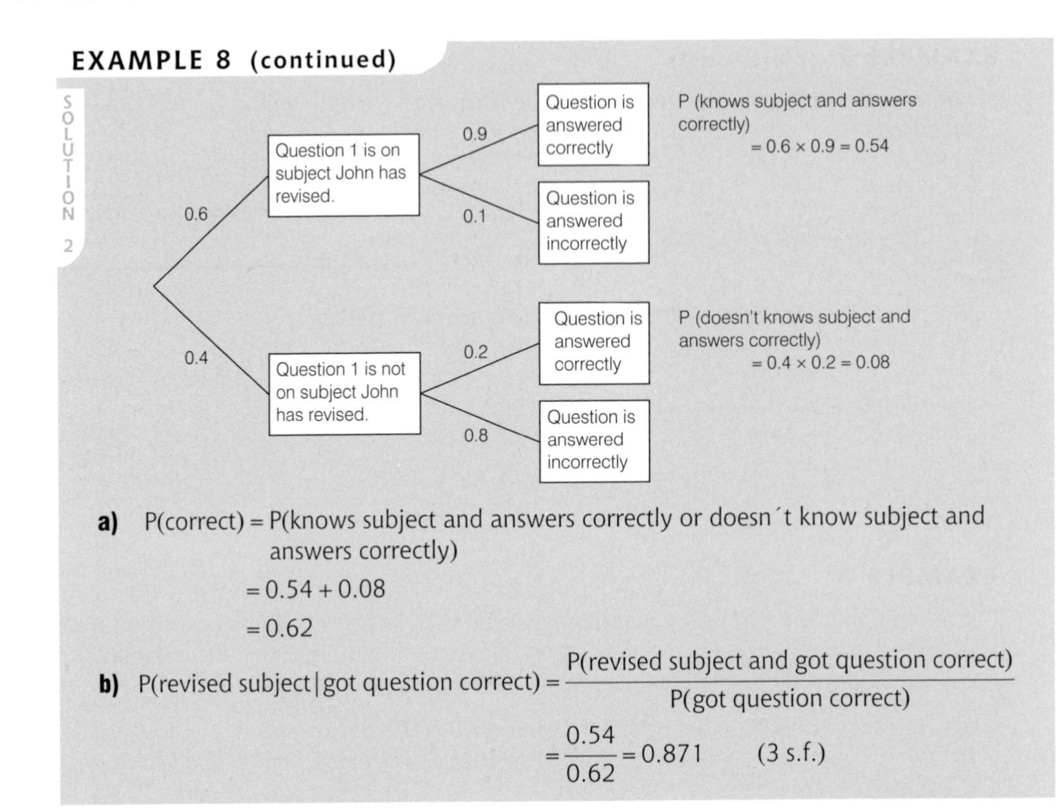

a) P(correct) = P(knows subject and answers correctly or doesn´t know subject and answers correctly)

$= 0.54 + 0.08$

$= 0.62$

b) $\text{P(revised subject} \mid \text{got question correct)} = \dfrac{\text{P(revised subject and got question correct)}}{\text{P(got question correct)}}$

$= \dfrac{0.54}{0.62} = 0.871 \quad \text{(3 s.f.)}$

EXAMPLE 9

A bag contains 4 red, 7 yellow and 9 green marbles. Three of the marbles are taken at random out of the bag.

a) Find the probability of getting 3 yellows.
b) Find the probability of getting 3 marbles the same colour.
c) Find the probability that the 3 balls are all yellow given that they are all the same colour.

SOLUTION 1

a) $\text{P(3 yellow)} = \text{P}(Y_1Y_2Y_3) = \frac{7}{20} \times \frac{6}{19} \times \frac{5}{18} = \frac{7}{228} = 0.030701\ ... = 0.0307 \quad \text{(3 s.f.)}$

b) P(same colour) = P(YYY or RRR or GGG)

$= 0.030701\ ... + \frac{4}{20} \times \frac{3}{19} \times \frac{2}{18} + \frac{9}{20} \times \frac{8}{19} \times \frac{7}{18}$

$= 0.10789\ ...$

$= 0.108 \quad \text{(3 s.f.)}$

c) $\text{P(3 yellow} \mid \text{all 3 are same colour)} = \dfrac{\text{P(3 yellow and all 3 are same colour)}}{\text{P(all 3 are same colour)}}$

$= \dfrac{\text{P(3 yellow)}}{\text{P(all 3 are same colour)}} = \dfrac{0.030701\ ...}{0.10789\ ...}$

$= 0.2845\ ...$

$= 0.285 \quad \text{(3 s.f.)}$

EXAMPLE 9 (continued)

SOLUTION 2

Again, a tree diagram could be used to answer this question.

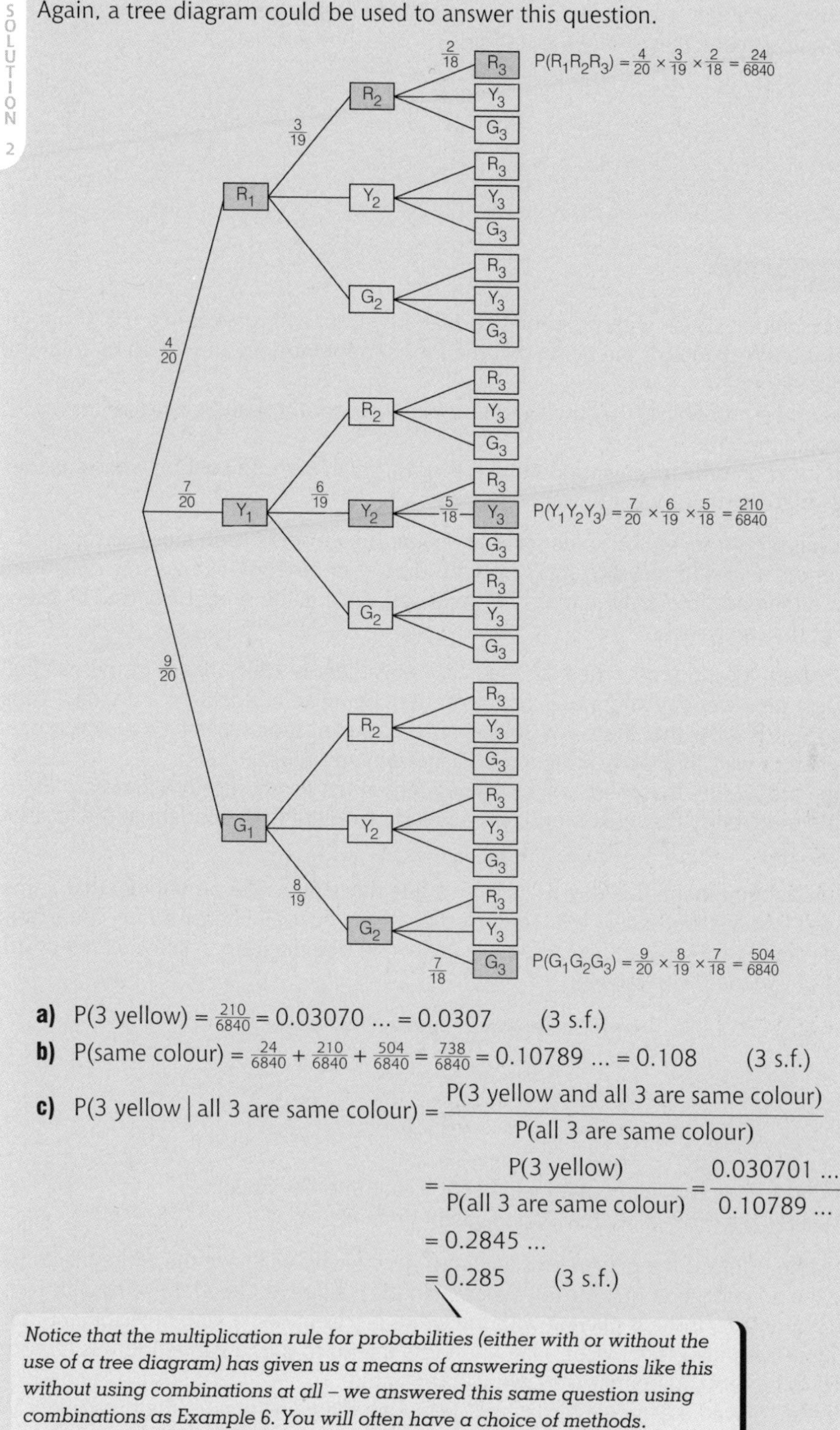

a) P(3 yellow) $= \frac{210}{6840} = 0.03070 \ldots = 0.0307$ (3 s.f.)

b) P(same colour) $= \frac{24}{6840} + \frac{210}{6840} + \frac{504}{6840} = \frac{738}{6840} = 0.10789 \ldots = 0.108$ (3 s.f.)

c) $$\text{P(3 yellow} \mid \text{all 3 are same colour)} = \frac{\text{P(3 yellow and all 3 are same colour)}}{\text{P(all 3 are same colour)}}$$

$$= \frac{\text{P(3 yellow)}}{\text{P(all 3 are same colour)}} = \frac{0.030701 \ldots}{0.10789 \ldots}$$

$$= 0.2845 \ldots$$

$$= 0.285 \quad \text{(3 s.f.)}$$

Notice that the multiplication rule for probabilities (either with or without the use of a tree diagram) has given us a means of answering questions like this without using combinations at all – we answered this same question using combinations as Example 6. You will often have a choice of methods.

EXAMPLE 10

Find the probability of getting 2 boys and 1 girl if 3 children are picked at random from a group containing 5 boys and 8 girls.

SOLUTION

$$P(\text{2 boys and 1 girl}) = P(B_1B_2G_3 \text{ or } B_1G_2B_3 \text{ or } G_1B_2B_3)$$
$$= \tfrac{5}{13}\times\tfrac{4}{12}\times\tfrac{8}{11}+\tfrac{5}{13}\times\tfrac{8}{12}\times\tfrac{4}{11}+\tfrac{8}{13}\times\tfrac{5}{12}\times\tfrac{4}{11}$$
$$= 3\times\tfrac{5}{13}\times\tfrac{4}{12}\times\tfrac{8}{11}=\tfrac{40}{143}=0.280 \quad \text{(3 d.p.)}$$

EXERCISE 4

1 I go to London by car with probability 0.8 and by train with probability 0.2. If I go by car there is a 0.6 chance of me being on time for my appointment and if I go by train this probability is 0.4.

a) Find the probability that the next time I go to London I shall be on time for my appointment.

b) If I was on time for my appointment the last time I went to London what was the probability that I went by car?

2 John has six red socks, four blue socks and eight brown socks. One morning during a power cut he has to pick two socks from his drawer at random. Draw a tree diagram to show all the possible combinations he could get. What is the probability that he gets a pair of matching socks?

3 At a certain location renowned for road accidents there is a 0.6 chance of there being an accident on a wet day and a 0.2 chance of there being an accident on a dry day. Long-term records show that there is a 30% chance of the location having a day which is classified as wet. By first drawing a tree diagram, determine:

a) the probability that there will be an accident at the location next Monday;

b) the probability that next Monday is a wet day if there is an accident at the location that day.

4 The probability that a fine day will follow a fine day is 3/5. The probability that a fine day will follow a wet day is 1/3. The first day, of my holiday Friday, is fine. If my holiday is four days long, copy and complete the following tree diagram to show all the possible types of weather I could have:

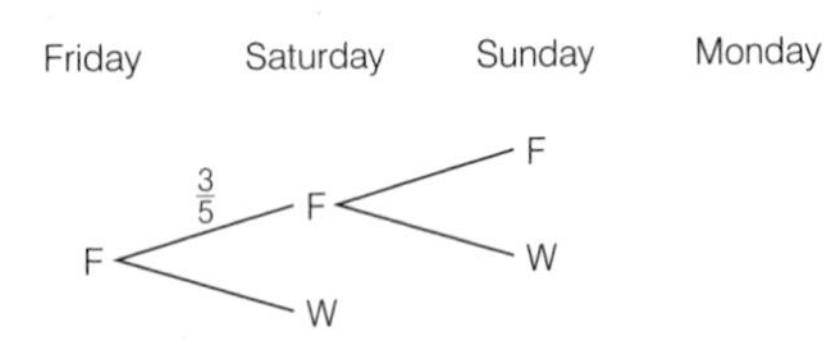

a) What is the probability that I will get at least three fine days?

b) What is the probability that Monday will be fine?

5 In a game of Chess the players toss a coin to decide who will have the white pieces (and have the advantage of first move). When Jean plays Clare at Chess, she wins with probability 0.75 if she is playing with the white pieces but only with probability 0.48 if she plays with the black pieces. Jean and Clare intend to play a game of Chess tomorrow.

a) Find the probability that Jean will win the game.

b) If they played a game yesterday and Jean won the game what is the probability that she was playing with the white pieces?

6 In a large group of people there are 35% males. 4% of the males are colour blind and 1% of the females are colour blind. If a person is chosen at random from this group find the probability that the person is:
a) male and colour blind;
b) colour blind;
c) male given that the chosen person is colour blind.

7 A certain disease is present in 3% of the population. There is a test to detect the presence of the disease, but unfortunately this test is not completely reliable. If a person has the disease there is a 98% chance that the test will give a positive result and if a person has not got the disease there is an 8% chance of the test giving a positive result.
A person is chosen at random from the population. Find the probability that:
a) the person has the disease and gives a positive response to the test;
b) the person gives a positive response to the test;
c) the person gives a negative response to the test;
d) the person does not have the disease given that they have given a positive response to the test.
Should this test be used? Explain your answer.

8 In a class of 25 children, 10 walk to school, 8 cycle and 7 are driven. If 3 children are selected at random from the class find the probability that:
a) they all walk;
b) they all travel by the same means;
c) one walks, one cycles and one is driven;
d) at least one walks to school;
e) one walks, one cycles and one is driven given that at least one walks.

9 A four-digit number is to be made by choosing four different digits from the set {1, 2, 3, 4, 5, 6, 7}. Find the probability that:
a) the number is even;
b) the number is greater than 6000;
c) the number is even and greater than 6000;
d) the number is even given that it is greater than 6000.

10 Mass produced glass bricks are inspected for defects. The probability that a brick has air bubbles is 0.002. If a brick has air bubbles then the probability that it is cracked is 0.5. If a brick is free from air bubbles then there is a probability of 0.005 of the brick being cracked. Calculate the probability that a randomly chosen brick:
a) has air bubbles and is cracked;
b) is cracked;
c) has air bubbles or is cracked;
d) has air bubbles given that it is cracked.

11 A and B play a series of exactly three games of Snooker. A has a 0.6 chance of winning the first game. If A wins a game he has a 0.7 chance of winning the next game. If B wins a game he has a 0.9 chance of winning the next game. Find the probability that:
a) A wins all three games;
b) A wins the series;
c) A won the first game given that he won the series.

12 Shinya and Hiroki take it in turns, starting with Shinya, to take a card, without replacement, from a pack of 10 cards containing the numbers 1, 2, 2, 3, 3, 3, 4, 4, 4, 4.
The first player to select a "4" card is the winner.
Find the probability that Hiroki wins this game.

Independence

Consider the experiment of throwing a pair of fair cubical dice. Let A and B be the events

A = "the score on the red die is 3"
B = "the score on the blue die is 2"

You know that P(A) = 1/6 and P(B) = 1/6.

Since the outcome of the red die doesn't affect the outcome of the blue die, you also know that $P(B \mid A) = 1/6$. It is said that the event B is **independent** of the event A.

Definition The events A and B are said to be INDEPENDENT if $P(B \mid A) = P(B)$.
(Or, in words, the fact that A has happened doesn't affect the likelihood of B happening.)

Using the multiplication rule, we see that if A and B are independent then

$$P(A \text{ and } B) = P(A) \times P(B \mid A) = P(A) \times P(B)$$

The result P(A and B) = P(A) × P(B) is only true if A and B are independent events and cannot be used in other circumstances.

Rule 6
If A and B are independent events then

$$\mathbf{P(A \text{ and } B) = P(A) \times P(B)}$$

Take care to only use this result when you know the events are independent!

EXAMPLE 11

John and Alan play three games of Chess. The probability of John winning any game is 0.6, the probability of a draw is 0.1, and the probability of Alan winning is 0.3. Find the probability that the series of three games:

a) is won by John;
b) is drawn;

and state any assumptions you have made in answering this question.

SOLUTION

a) P(J wins) = P(JJJ or JJD or JDJ or DJJ or JJA or JAJ or AJJ or JDD or DJD or DDJ)

$$= 0.6^3 + 3 \times 0.6^2 \times 0.1 + 3 \times 0.6^2 \times 0.3 + 3 \times 0.6 \times 0.1^2 = 0.648$$

b) P(draw) = P(DDD or DJA or DAJ or ADJ or AJD or JDA or JAD)

$$= 0.1^3 + 6 \times 0.1 \times 0.6 \times 0.3 = 0.109$$

It is assumed that the result of each game is independent of the results of the previous games.

EXAMPLE 12

John has three cubical dice. Each has three faces coloured red, two faces coloured green and one face coloured yellow. John throws the three dice. Find the probability that he gets:

a) the same colour on all three dice;
b) a different colour on each dice;
c) just two different colours.

EXAMPLE 12 (continued)

SOLUTION

a) P(same colour) = P(RRR or GGG or YYY) = $(\frac{3}{6})^3 + (\frac{2}{6})^3 + (\frac{1}{6})^3 = \frac{1}{6}$

b) P(different colours) = 6 × P($R_1 G_2 Y_3$) = $6 \times \frac{3}{6} \times \frac{2}{6} \times \frac{1}{6} = \frac{1}{6}$

c) P(two colours) = 1 – P(all same OR all different) = $1 - [\frac{1}{6} + \frac{1}{6}] = \frac{2}{3}$

EXERCISE 5

1 David throws a fair cubical die and then picks a card at random from a standard pack. Find the probability that he gets:

a) a 5 on the die and a diamond;
b) a 1 on the die and an ace;
c) an even number on the die and a card which isn't a diamond.

2 The probability that a person likes chips is $\frac{4}{5}$. The probability that a person likes cream cakes is $\frac{2}{3}$. Assuming that these two events are independent, find the probability that a randomly chosen person:

a) likes chips and cream cakes;
b) likes chips and doesn't like cream cakes;
c) doesn't like chips and doesn't like cream cakes.

3 Next Saturday, the probability that James will go to the cinema is $\frac{3}{5}$ and the probability that Karen will go to the cinema is $\frac{1}{3}$. Assuming that these two events are independent, find the probability that:

a) both James and Karen will go to the cinema;
b) neither James nor Karen will go to the cinema.

4 When I drive home there is a 0.7 chance of me being stopped at a set of traffic lights.

a) What is the probability that I will be stopped on my way home on both Monday and Tuesday?
b) What is the probability that I will be stopped on Wednesday but not on Thursday?

5 When Alan and Bill play Chess, Alan wins with probability 4/9, Bill wins with probability 1/3, and the probability of a draw is 2/9. They decide to play a series of exactly three games. Find the probability that:

a) Alan wins all three games;
b) Bill wins at least one game;
c) Bill wins the series;
d) the series is drawn.

6 Two teams, A and M, play a football match against each other. The numbers of goals each scores in the match are independently distributed, with the distributions shown below:

Number of goals	0	1	2	3
Probability of A scoring this number of goals	0.1	0.3	0.3	0.3
Probability of M scoring this number of goals	0.2	0.3	0.4	0.1

a) Prove that the probability that A wins a game is 0.48.
b) Find the probability that M wins a game.
c) Find the probability that a game is drawn.

A victory in a game is worth 3 points to the winning team and 0 points to the losing team; a drawn game is worth 1 point to each team.
If the two teams play each other twice in a season, find the probability that A scores more points than M from the two matches.

7 An unbiased blue die has faces marked 1, 2, 3, 4, 5, 6. Another unbiased red die has faces marked 1, 2, 2, 3, 3, 3. In a game a card is selected at random from a standard pack of 52 playing cards and if a diamond is obtained the blue die is thrown and otherwise the red die is thrown. Find the probability that:

a) the score on the die is 3;

b) the red die was thrown given that the score obtained was a 3.

8 An e-mail sent to company C has probability 0.6 of being answered on the next day, probability 0.3 of being answered on the second day after it was sent, and probability 0.1 of being answered on the third day. An e-mail sent to company K has corresponding probabilities of 0.2, 0.5 and 0.3. Two e-mails are sent by the same person at the same time, one to each company. Assuming that the answering times of C and K are independent, calculate the probabilities that:

a) at least one e-mail will be answered the next day;

b) the e-mail sent to C will be answered one day earlier than the e-mail sent to K;

c) both e-mails will be answered on the same day.

9 Two poor marksmen, A and B, fight a duel. The probability that A will hit B on any attempt is 1/5. The probability that B will hit A on any attempt is 1/3. A shoots first and then B and A shoot alternately until someone is hit or until A and B have used up their stock of 3 bullets each. Find the probability that:

a) A hits B;

b) A is not hit.

10 Three baskets each contain 10 identically sized balls. The first basket contains 7 red and 3 yellow balls; the second basket contains 5 red and 5 yellow balls, and the third basket contains 2 red and 8 yellow balls. James picks one ball at random from each basket. Find the probability that he obtains:

a) three red balls;

b) at least one yellow ball;

c) two red and one yellow ball;

d) at least one ball of each colour.

11 Three men, U, V and W, share an office with a single telephone. Calls come in at random in the proportions 3/7 for U, 3/7 for V and 1/7 for W. Their work means that the men have to leave the office at random times, so that U is out for half his working time and V and W are each out for a quarter of their time.

Find the probability that the next phone call to the office:

a) is for U and he is able to answer it;

b) can be answered by the required person;

c) cannot be answered since there is no-one in.

Find the probability that the next three calls:

d) are all for the same man;

e) are for three different men.

A caller wishes to obtain V. Find the probability that he has to make more than two attempts to successfully get V.

12 A box contains a large number of three types of packets of biscuits: type A packets contain 20 milk chocolate biscuits; type B contain 15 plain chocolate biscuits; type C contain 12 milk chocolate biscuits and 8 plain chocolate biscuits. The number of packets of the three types are in the ratio 2 : 3 : 4.

A packet is drawn at random and a biscuit selected at random from it.
a) What is the probability that this biscuit is a milk chocolate biscuit?
A second biscuit is now drawn at random from those remaining in the chosen packet. Find the probability that:
b) both biscuits chosen are milk chocolate biscuits;
c) the second biscuit chosen is a milk chocolate biscuit;
d) the second biscuit chosen is a milk chocolate biscuit given that the first chosen was a plain chocolate biscuit.

Having studied this chapter you should

- understand the terms "exclusive events"; "conditional probability" and "independent events"
- know and be able to use the laws of probability:
  (1) P(A) = 1 − P(not A)
  (2) P(A or B) = P(A) + P(B) if A and B are exclusive events
  (3) P(A or B) = P(A) + P(B) − P(A and B)
  (4) $\mathrm{P(A|B)} = \dfrac{\mathrm{P(A\ and\ B)}}{\mathrm{P(B)}}$
  (5) P(C and D) = P(C) × P(D | C)
  (6) P(A and B) = P(A) × P(B) if A and B are independent events

REVISION EXERCISE

1 Aamir, Brad and Chris each take a penalty shot in a sports competition. The probability that Aamir succeeds is $\frac{3}{4}$, the probability that Brad succeeds is $\frac{2}{5}$ and the probability that Chris succeeds is $\frac{5}{9}$. The result of any competitor's shot is independent of all other results. Calculate the probability that:
i) Aamir, Brad and Chris all succeed;
ii) Aamir and Brad succeed but Chris does not;
iii) exactly two out of the three competitors succeed.

(OCR Nov 2002 S1)

2 Benjamin and Winston play three games of table tennis against each other. The probability that Benjamin wins the first game is 0.7. For each subsequent game, the probability of a player winning the game is 0.6 if that player won the previous game. (A table tennis game cannot be drawn.)
i) Draw a tree diagram to show this information.
ii) Use the tree diagram to find the probability that

a) Winston wins all three games;
b) Winston wins one game and Benjamin wins two games.

3 A cylinder contains 3 green, 4 red and 5 yellow marbles. Two marbles are removed at random from the cylinder, one after the other. Once a marble has been removed it is not replaced in the cylinder. Find the probability that:

a) the first marble is red;
b) the second marble is yellow if the first marble is green;
c) the first marble is green and the second marble is yellow;
d) the second marble is yellow;
e) at least one marble is yellow.

4 A lady decides to pick one flower at random from each of two flower beds. The first flower bed has 6 red flowers, 4 yellow flowers and 2 white flowers; the second flower bed has 8 yellow flowers and 12 white flowers.

a) Calculate the probability that the flower chosen from the first bed is yellow.
b) Calculate the probability that both the flowers chosen are yellow.
c) Calculate the probability that both flowers chosen are the same colour.
d) Calculate the probability that the flowers chosen are of different colours.

5 A class contains 24 pupils. 12 of the pupils travel to school by bus, 7 travel by car and the remainder walk to school. If 3 pupils are chosen at random from the class, find the probability that:

a) two travel by bus and one by car;
b) one travels by bus, one by car and one walks;
c) at most two travel by bus.

If all 3 pupils travel to school by the same means, what is the probability that they all travel by bus?

6 Three fair cubical dice, each having faces marked 1, 2, 3, 4, 5 and 6, are thrown. Find the probability that the total score is at least 6.

7 Jonathan has two bags of balls. The first bag contains 4 red and 4 blue balls. The second bag contains 2 green and 6 yellow balls. He takes two balls at random and without replacement from each bag. Find the probability that:

a) he takes two red balls from the first bag;
b) he takes two yellow balls from the second bag;
c) he takes two red balls from the first bag and two yellow balls from the second bag;
d) he takes two red balls from the first bag or two yellow balls from the second bag.

8 Elizabeth has probability 0.7 of passing her history exam and probability 0.9 of passing her science exam. Find the probability that:

a) she passes both exams;
b) she passes just one of these exams.

State any assumptions you have made in answering this question.

9 A jar contains 4 red discs, 6 white discs and 5 green discs. Three discs are removed at random from the jar, one after the other. Once a disc is removed it is not replaced in the jar.

a) Find the probability that:

i) the first disc is red;
ii) the second disc is white if the first disc is green.

b) Calculate the probability that:

i) exactly two of the three discs are red;
ii) the three discs are the same colour.

(OCR May 2002 S1)

10 Students have to take an examination in Mathematics and an examination in Geography. Students are not allowed to take a subject more than twice, and it is to be assumed that performance in one subject does not affect performance in the other subject. For a randomly chosen student, the probability of passing Mathematics at the first attempt is 0.8. If a student fails, a second attempt is required, and then the probability of passing is 0.9. This information is shown in the tree diagram below:

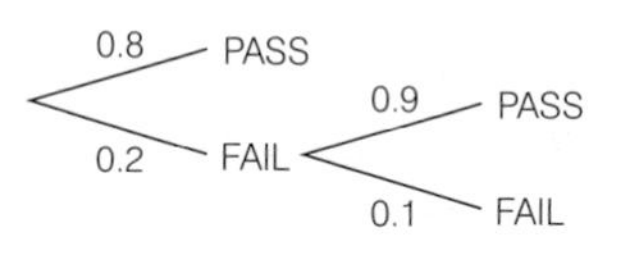

i) Calculate the probability that a randomly chosen student fails Mathematics at both attempts.

ii) Given that a pass in Mathematics is achieved, calculate the probability that it is obainted at the second attempt, giving your answer correct to 3 significant figures.

In Geography the probability, for a randomly chosen student, of passing at the first attempt is 0.7. If a student fails, a second attempt is required, and then the probability of passing is 0.9. Calculate the probability that a randomly chosen student:

iii) passes both subjects at the first attempt;

iv) fails both subject twice;

v) has to retake one examination, but achieves two passes.

(OCR Jun 1996 P1)

6 Further Counting Methods

The purpose of this chapter is to enable you to

- revise the multiplication, permutation and combination rules introduced in Chapter 4
- use appropriate methods to calculate the number of arrangements when restrictions are imposed
- calculate the number of different ways of arranging items which are not all distinct

The Multiplication Rule, the Permutation Rule and the Combination Rule

When first looking at counting methods (Chapter 4) three basic results were established:

- **Multiplication Rule**
  If there are m ways of doing a first task and n ways of doing a second task then there are mn ways of doing the two tasks.
- **Permutation Rule**
  The number of ways of choosing r items from a set of n different items when the order of choice is taken into account is

  $${}_n\mathrm{P}_r = n(n-1)(n-2)\ldots(n-r+1) = \frac{n!}{(n-r)!}$$

  [In this case "abc" is a different choice from "bac".]
- **Combination Rule**
  The number of ways of choosing r items from a set of n different items when the order of choice is not to be taken into account is

  $$\binom{n}{r} = {}_n\mathrm{C}_r = \frac{n!}{(n-r)!r!}$$

  [In this case "abc" is regarded as the same choice as "bac".]

This chapter revises the usage of these three results and extends their application to more complicated situations.

EXAMPLE 1

A class contains 25 boys.

a) In how many ways can a five-a-side football team of 5 players be chosen from the class?
b) In how many ways can a group of six boys be chosen from the class to go on a trip to the cinema?

EXAMPLE 1 (continued)

SOLUTION

a) In this case the order of choice is important since a team having David as goalkeeper and Michael as centre-forward is different from one having Michael as goalkeeper and David as centre-forward. Permutations should therefore be used.

Number of possible teams = ${}_{25}P_5 = 6\ 375\ 600$

b) In this case the order of choice does not matter so combinations can be used.

Number of possible choices = ${}_{25}C_6 = 177\ 100$

EXAMPLE 2

A bookshelf contains 4 paperback books and 7 hardback books. James selects five books at random from the bookshelf.

a) In how many different ways can James select the five books?
b) In how many different ways can James make a selection which has exactly two paperbacks?
c) Calculate the probability that James chooses exactly two paperbacks, stating any assumption that needs to be made in order to calculate the probability.

SOLUTION

The order of choice of the books is unimportant, so combinations will be used.

a) Number of possible selections ${}_{11}C_5 = 462$
b) If James chooses two paperbacks then he must also choose three hardbacks. There are ${}_4C_2$ ways of choosing two paperbacks from the four paperbacks and there are ${}_7C_3$ ways of choosing three hardbacks from the seven hardbacks. The multiplication rule now implies that:
Number of ways of choosing 2 paperbacks and 3 hardbacks = ${}_4C_2 \times {}_7C_3 = 210$
c) Provided each possible selection of books is equally likely,

$$P(2\text{ paperbacks}) = \frac{210}{462} = \frac{5}{11}.$$

EXAMPLE 3

Ahmed creates a four-letter string by choosing four letters at random from the alphabet and placing them in a line. He allows himself to repeat letters. For example, if he chooses the letters **E, X, T** and **E** then his final string is "EXTE".

a) How many different strings can Ahmed produce in this way?
b) How many of these strings consist of four different letters?
c) What is the probability that at least one letter is repeated in Ahmed's string?

SOLUTION

a) Ahmed has 26 choices for each letter of his string, so

Number of possible strings = $26 \times 26 \times 26 \times 26 = 26^4 = 456\ 976$

b) The number of strings with four different letters is the number of ways of choosing 4 letters from 26 when the order matters, so

Number of strings with 4 different letters = ${}_{26}P_4 = 358\ 800$

EXAMPLE 3 (continued)

c) Using the answers to (b) and (c) gives

$$\text{P(string has 4 different letters)} = \frac{358\ 800}{456\ 976} = 0.7852 \qquad \text{(4 d.p.)}$$

so

$$\begin{aligned}\text{P(at least 1 letter repeated)} &= 1 - \text{P(string has 4 different letters)}\\ &= 1 - 0.7852\\ &= 0.2148 \qquad \text{(4 d.p.)}\end{aligned}$$

EXAMPLE 4

A group of eight people includes Mr and Mrs Smith. In how many ways can a group of four people be chosen from this group in such a way that Mr and Mrs Smith are not both in the group?

SOLUTION

Number of groups with no members of the Smith family = ${}_6C_4 = 15$

Number of groups with just one member of the Smith family = ${}_2C_1 \times {}_6C_3 = 40$

So number of groups without both Mr and Mrs Smith = 15 + 40 = 55

No. of ways of choosing one member of Smith family = ${}_2C_1$

No. of ways of choosing three other people = ${}_6C_3$

EXERCISE 1

Take care to determine whether permutations or combinations should be used in these questions.

1 In how many ways can 5 boys be chosen from a class of 20 to move chairs into the assembly hall?

2 A carol service is to have 6 different readings. The readers are to be selected from a class of 15 girls. In how many different ways can the selection be made?

3 Four numbers are to be chosen from the set {2, 3, 6, 8, 9} and put together to form a four-digit number. How many different four-digit numbers can be put together if:
a) there is to be no repetition of digits;
b) repetition of digits is allowed.

4 A class contains 10 boys and 15 girls. A group of 6 pupils is to be selected from the class to go on a theatre visit. In how many ways can this be done if:
a) any combination is allowed;
b) the group must contain two boys and four girls;
c) the group must contain pupils of just one sex;
d) the group must contain at least two boys and at least two girls.

5 If a hand of eight cards is dealt from a standard pack, find the probability that:
a) the hand consists of six spades and two hearts;
b) the hand consists of six cards of one suit and two cards of another suit.

6 The diagram shows the seating plan in a small minibus which has eight passenger seats plus a seat for the driver. A group of eight friends hire the minibus for an outing to the seaside. Only four of the eight are allowed to drive the minibus.
In how many different ways can the eight friends sit in the minibus as they travel to the seaside?

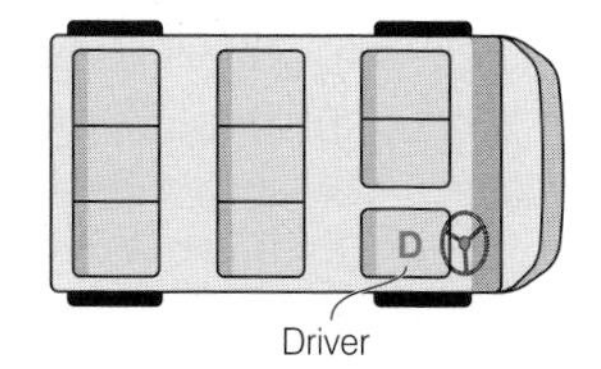

7 A class contains 10 boys and 15 girls. At the end of the school year, 8 of the pupils can go on a visit to a local castle, 7 of the pupils can go to the theatre and the remaining 10 will go to a museum. In how many ways can the class be divided for their activity day:
a) if there are no restrictions;
b) if the group going to the theatre must consist of 4 girls and 3 boys?

8 A bag of sweets contains 12 chocolates and 8 toffees. If Nick picks 5 sweets at random from the bag find the probability that he will obtain:
a) exactly 2 toffees;
b) at least 2 chocolates and at least 2 toffees.

9 A group of 12 people consists of Mr and Mrs Edmunds, Mr and Mrs Jones, Mr and Mrs Hartley, Mr and Mrs McGladdery, Mr and Mrs Harrison and Mr and Mrs Walker. In how many different ways can:
a) four people be chosen from this group of 12;
b) two married couples be chosen from this group of 12;
c) four people be chosen if Mr and Mrs Jones cannot both be chosen?

Conditional Arrangements

You have seen how to calculate the number of ways a group of four people can be chosen from a group of eight and, in Example 4, we have also seen that it is possible to calculate the number of possible combinations when restrictions are imposed on the choice. This section considers some further examples where the choice is restricted.

EXAMPLE 5

A group of 10 friends, including Alan, Brian, Colin, David and Edward, are to stand in a line. In how many ways can this be done:

a) if there are no restrictions;
b) if Alan, Brian and Colin must stand next to each other;
c) if David and Edward must not stand next to each other?

SOLUTION

a) There are 10! = 3 628 800 ways of arranging the 10 friends.
b) If you call the friends A, B, C, D, E, F, G, H, I and J then you have got to keep A, B and C grouped together.

EXAMPLE 5 (continued)

You then have the group of three as a single unit and seven individuals to arrange in a line. In all, there are **eight** items to arrange in a line: this can be done in $8! = 40\,320$ ways.
The group of three can then be arranged in $3! = 6$ different ways.
The multiplication rule means that there are $40\,320 \times 6 = 241\,920$ ways of arranging the 10 friends in a such a way that Alan, Brian and Colin are all together.

c) It is easiest to first find the number of arrangements where David and Edward do stand next to each other.

Counting D and E as a single unit, there are **nine** items to arrange in the line.
The group consisting of D and E can be arranged in 2! ways, so
Number of arrangements with D and E next to each other $= 9! \times 2! = 725\,760$
Number of arrangements with D and E **not** next to each other

$= 3\,628\,800 - 725\,760$
$= 2\,903\,040$

EXAMPLE 6

a) In how many different ways can a group of five men and five women be arranged in a line so that the sexes alternate along the line.

b) Ollie has 13 letter tiles with the letters

A, E, I, O, U, B, C, D, F, G, H, J, K

written on them. In how many ways can he arrange all these tiles in a straight line in such a way that no two vowels are next to each other?

SOLUTION

a) Start by picturing the line where the people must be arranged:

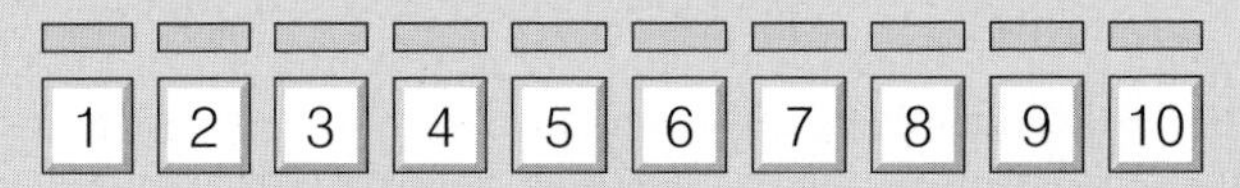

If you decide to put men in positions 1, 3, 5, 7 and 9 then there are 5! ways of arranging the five men; similarly there are 5! ways of arranging the five women in positions 2, 4, 6, 8 and 10.
The multiplication rule tells you that there are $5! \times 5!$ ways of arranging the men into the odd numbered positions and the women into the even numbered positions. However, you could have the women in the odd numbered positions and the men in the even numbered positions, so

Total number of arrangements $= 2 \times 5! \times 5! = 28\,800$

b) Start by placing the eight consonants into a line. This can be done in 8! different ways. Having placed the consonants, you can see from the diagram that there are nine possible places to put a vowel:

EXAMPLE 6 (continued)

You have got to choose five of these possible spaces to place a vowel into. This can be done in ${}_9P_5 = 9 \times 8 \times 7 \times 6 \times 5$ different ways.

No. of possible arrangements $= 8! \times {}_9P_5 = 609\ 638\ 400$

Permutations where Not All the Objects are Different

EXAMPLE 7

How many different 7-letter strings can be made using each of the letters from the word "BANANAS" exactly once?

SOLUTION

Start by distinguishing between the different "A"s and the different "B"s.

You know that the letters of "$BA_1N_1A_2N_2A_3S$" can be arranged in $7! = 5040$ different ways.

Now A_1 A_2 A_3 can be arranged in 3! different ways so each different arrangement of "BANANAS" will be repeated 3! times in the 7! arrangements of "$BA_1N_1A_2N_2A_3S$".

The arrangements

$BA_1N_1A_2N_2A_3S$	$BA_1N_1A_3N_2A_2S$
$BA_2N_1A_1N_2A_3S$	$BA_2N_1A_3N_2A_1S$
$BA_3N_1A_1N_2A_2S$	$BA_3N_1A_2N_2A_1S$

are all indistinguishable once the subscripts are dropped.

Similarly, N_1 N_2 can be arranged in 2! different ways so each different arrangement of "BANANAS" will be repeated 2! times in the 7! arrangements of "$BA_1N_1A_2N_2A_3S$".

The arrangements

$BA_1N_1A_2N_2A_3S$ $\quad$ $BA_1N_2A_2N_1A_3S$

are indistinguishable from each other once the subscripts are dropped.

The number of **different** arrangements of "BANANAS" must therefore be

$$\frac{7!}{3! \times 2!} = \frac{5040}{6 \times 2} = 420$$

This argument can be generalised to give

The number of arrangements of n items when k_1 are alike of a first kind, k_2 are alike of a second kind, k_3 are alike of a third kind, ... is

$$\frac{n!}{k_1!k_2!k_3!\ ...}$$

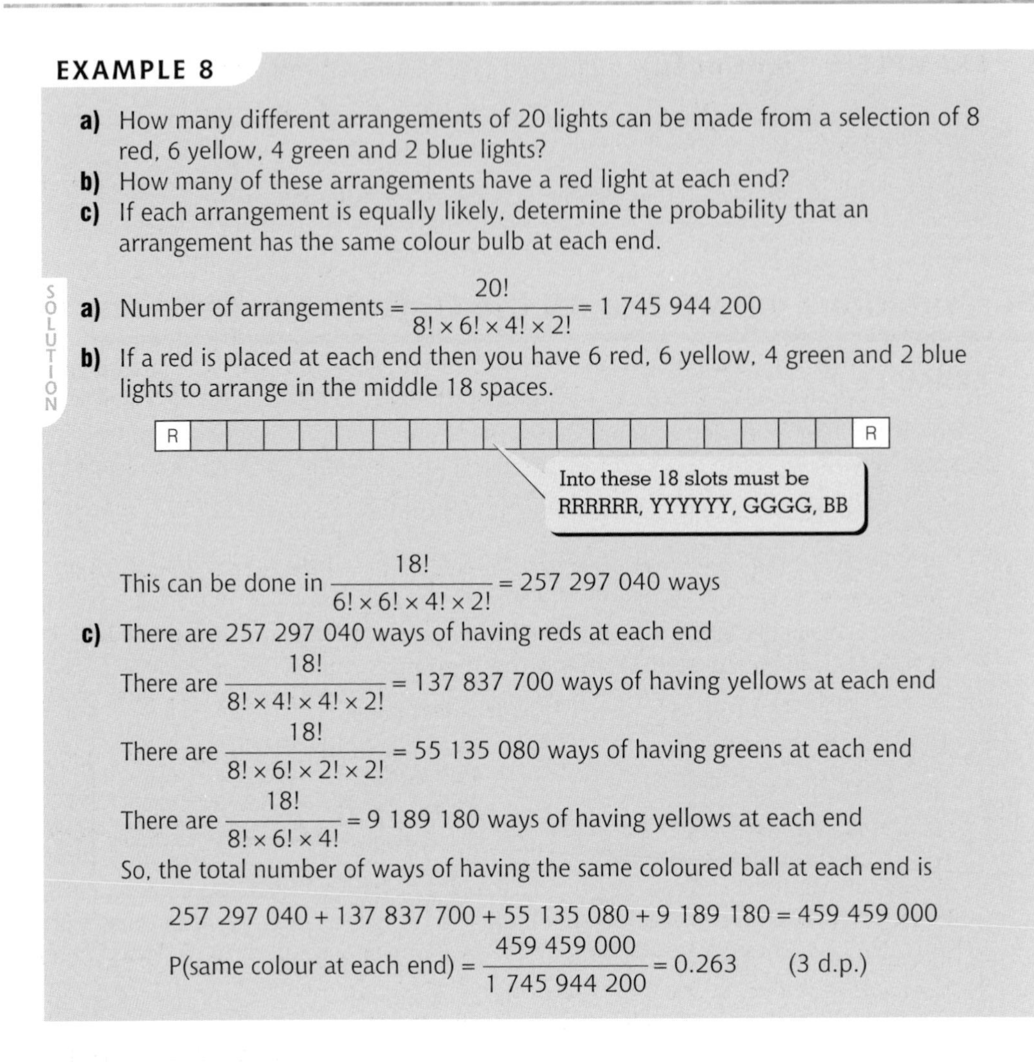

EXAMPLE 8

a) How many different arrangements of 20 lights can be made from a selection of 8 red, 6 yellow, 4 green and 2 blue lights?
b) How many of these arrangements have a red light at each end?
c) If each arrangement is equally likely, determine the probability that an arrangement has the same colour bulb at each end.

SOLUTION

a) Number of arrangements $= \dfrac{20!}{8! \times 6! \times 4! \times 2!} = 1\ 745\ 944\ 200$

b) If a red is placed at each end then you have 6 red, 6 yellow, 4 green and 2 blue lights to arrange in the middle 18 spaces.

This can be done in $\dfrac{18!}{6! \times 6! \times 4! \times 2!} = 257\ 297\ 040$ ways

c) There are 257 297 040 ways of having reds at each end

There are $\dfrac{18!}{8! \times 4! \times 4! \times 2!} = 137\ 837\ 700$ ways of having yellows at each end

There are $\dfrac{18!}{8! \times 6! \times 2! \times 2!} = 55\ 135\ 080$ ways of having greens at each end

There are $\dfrac{18!}{8! \times 6! \times 4!} = 9\ 189\ 180$ ways of having yellows at each end

So, the total number of ways of having the same coloured ball at each end is

$$257\ 297\ 040 + 137\ 837\ 700 + 55\ 135\ 080 + 9\ 189\ 180 = 459\ 459\ 000$$

$$\text{P(same colour at each end)} = \frac{459\ 459\ 000}{1\ 745\ 944\ 200} = 0.263 \qquad \text{(3 d.p.)}$$

EXERCISE 2

1 A group of 8 friends, including Anne and Belinda, are to stand in a straight line. In how many ways can this be done if:
a) there are no restrictions;
b) Anne and Belinda must stand together;
c) Anne and Belinda must not stand together?

2 Anne, Brian, Cathy, David, Edwina and Frank are to stand together in a straight line. In how many ways can this be done if:
a) there are no restrictions;
b) Anne, Brian and Cathy must be next to each other;
c) David and Edwina must not be next to each other;
d) the line must go girl, boy, girl, boy, girl, boy;
e) the sexes must alternate along the line?

3 How many different 10-letter strings can be made using each of the letters of "STATISTICS" exactly once?

4 How many different eight-digit numbers can be made using each of the digits

3, 3, 3, 4, 5, 5, 5, 6

exactly once?

5 A group of eight people, including Mr and Mrs Edwards and Mr and Mrs Smith, are to stand in a line.

a) Calculate the number of different ways that this can be done.

b) Calculate the number of different ways in which this can be done if Mr and Mrs Smith must stand next to each other and Mr and Mrs Edwards must also stand next to each other.

c) Assuming that each possible arrangement is equally likely, what is the probability that Mr and Mrs Smith will be standing next to each other and that Mr and Mrs Edwards will be standing next to each other?

6 James has 12 flags: four are red, three are green, two are white and there are single yellow, blue and orange flags. How many different arrangements are there of these flags in a line?

7 The digits 1, 2, 3, 4, 5, 6, 7, 8, 9 are to be arranged at random to form a nine-digit number. In how many ways can this be done?

a) What is the probability that the resulting number is greater than 500 000 000?

b) What is the probability that the resulting number is even?

c) What is the probability that the even numbers are consecutive digits of the nine-digit number?

d) What is the probability that the 1 and 7 are not consecutive digits of the nine-digit number?

8 Three boys and four girls are to be arranged in a straight line for a photograph. In how many ways can this be done if:

a) the arrangement must be girl, boy, girl, boy, girl, boy, girl;

b) no two boys are next to each other?

9 Thomas has five red balls, numbered 1, 2, 3, 4 and 5, five yellow balls, numbered 1, 2, 3, 4 and 5, and five green balls, numbered 1, 2, 3, 4 and 5.

In how many ways can the balls be arranged in a line if:

a) all the red balls must be together, all the yellow balls must be together and all the green balls must be together;

b) all the 1 balls must be together, all the 2 balls must be together, all the 3 balls must be together, all the 4 balls must be together and all the 5 balls must be together;

c) all the red balls must be together;

d) all the red balls must be together and all the yellow balls must be together?

10 Mr and Mrs Jones and their son Ollie go to the theatre with five friends. The eight seats they have been assigned are in two rows with four adjacent seats in each row.

In how many different ways can the party of eight people sit if the Jones family insist on sitting next to each other in one row?

11 A bag of letter tiles contains 10 tiles each with one letter. The letters on the tiles are

m, a, n, c, h, e, s, t, e, r

Four tiles are to be selected from the bag but the order of choice is not important. The tiles are not replaced after being selected.

a) How many choices of four letters can be made which do not include an **e**?

b) How many choices can be made which have exactly one tile with an **e** on?

c) How many choices can be made which have two tiles with an **e** on?

How many possible choices are there altogether?

Having studied this chapter you should

- be able to use the multiplication, permutation and combination rules for simple arrangements and for examples involving conditional arrangements
- know that the number of arrangements of n items when k_1 are alike of a first kind, k_2 are alike of a second kind, k_3 are alike of a third kind, ... is

$$\frac{n!}{k_1!k_2!k_3!\ \ldots}$$

REVISION EXERCISE

1 Three married couples, Mr and Mrs Aziz, Mr and Mrs Baker and Mr and Mrs Campbell, are arranged in a line for a photograph.

i) How many different arrangements of the six people are possible?

ii) In how many of these arrangements is Mr Aziz standing next to his wife?

iii) Given that every possible arrangement is equally likely, calculate the probability that Mr Aziz is standing next to his wife.

(OCR May 2002 S1)

2 A football squad contains three goalkeepers, six defenders, five midfield players and seven forwards. A six-a-side team is to be selected from the squad by choosing a goalkeeper, two defenders, a midfield player and two forward players. How many different possible teams could be selected from the squad?

3 Mark has a bookshelf which holds 12 paperback books and 8 hardback books. He decides to take 4 books away on holiday with him.

a) How many different possible choices of four books are there?

b) In how many different possible ways can Mark choose three paperbacks and one hardback?

c) What is the probability that Mark takes at least three paperbacks on holiday if he chooses the books at random?

4 At a party there are 6 girls and 4 boys. The 10 people are to form a line for a dance.

a) In how many different ways can this be done?

Peter and Hayley are two of the people at the party.

b) In how many different ways can the line be formed if Peter and Hayley must be next to each other in the line?

In how many ways can the line be formed if:

c) all the boys must be together and all the girls must be together;

d) no two boys must be next to each other in the line?

5 Paul has 12 flags which are identical in size and shape. Three of the flags are red, four are blue and five are white. The 12 flags are displayed in a line.

a) How many different possible displays of the flags could be produced?

b) How many of these possible displays have a white flag at each end?

If Paul hangs the flags at random, calculate the probability that the flags at each end of the display are the same colour.

6 Michael has 10 tiles each with a single number written on it. The numbers on the tiles are

2, 2, 6, 6, 6, 7, 7, 7, 7, 8

Michael arranges the tiles in a line to produce a 10-digit number.

a) How many different 10-digit numbers can Michael produce?
b) How many of these 10-digit numbers start with **7 6**?
c) How many of these 10-digit numbers are greater than 7.5×10^9?

7 The board of directors of a company consists of 4 men and 4 women. The 8 directors are told to stand in a straight line so that a photograph can be taken.

i) Calculate the number of different ways in which the 8 directors can be arranged in a line.
ii) In how many ways can the directors be arranged so that in the line the men and women stand alternately?
iii) Five members of the board of directors are to be chosen to form a committee. There must be at least 2 women on the committee. Find the number of different possible committees that could be chosen.

(OCR Jan 2001 S1)

8 A bag contains 30 plastic tiles which are used in a word game. Each tile has a single letter written on it. 12 of the tiles have vowels written on them and the remaining 18 tiles have consonants written on them. A contestant in the game picks 7 tiles at random, without replacement.

i) Find the probability that, of the 7 tiles, 4 have vowels written on them and 3 have consonants written on them.
ii) Find the probability that, of the 7 tiles, at least 1 has a vowel written on it.
iii) The letters written on the tiles are **A B A E S S U**. Calculate the number of different arrangements of these letters if the tiles are to be placed in a straight line.

(OCR Jun 2001 S1)

9 Ten discs have a letter written on them. Five have A, three have B and two have C written on them. The discs are then placed in a bag and selected one at a time, at random. They are then placed in a straight line in the order in which they were selected. The sequence of 10 letters makes a code word. So, for example, one possible code word is ABBAAACBCA.

i) How many different possible code words can be made?
ii) How many different possible code words start and end with the letter B?
iii) Find the probability that a code word starts and finishes with the same letter.

(OCR Nov 2002 S1)

10 a) Find the number of different arrangements of all the letters of the word "papaya".
b) If three letters are chosen at random from the word "papaya", find the probability that at least two of them are "a"s.

7 Discrete Random Variables

The purpose of this chapter is to enable you to

- find the probability distribution of a discrete random variable
- calculate the expected value (or mean), variance and standard deviation of a discrete probability distribution

Probability Distributions

You have already met the idea of a frequency distribution for a set of quantitative data, which is simply a table listing all the observed data values together with the corresponding frequencies. This chapter introduces a similar idea: the probability distribution of a random variable.

A **random variable** is a random experiment whose sample space consists of numbers. Random variables are usually denoted by capital letters.

Examples of random variables include:

- U = the score when a standard cubical die is thrown. The sample space of U is the set $S_U = \{1, 2, 3, 4, 5, 6\}$
- V = the number of sixes obtained when a die is thrown 15 times. The sample space of V is the set $S_V = \{0, 1, 2, 3, ..., 14, 15\}$
- W = the number of times a coin must be tossed to obtain the first "head". The sample space of W is the set $S_W = \{0, 1, 2, 3, ...\}$
- T = the length of time, in minutes, that a passenger must wait on the platform of an underground railway station where the trains run regularly every 10 minutes. The sample space of T is the set of all numbers, including all decimals, that lie in the interval $0 \leqslant t < 10$.

The first three of these random variables are described as being **discrete** random variables since their sample spaces consist of separate values that have a definite gap between each other. The fourth random variable, T, is described as being a **continuous** random variable since its sample space consists of a whole interval of numbers. In this module only discrete random variables will be considered but continuous random variables are studied in S2.

The **probability distribution** of a discrete random variable is a table, or rule, listing all the possible values that the random variable can take together with the corresponding probabilities. Remember that the probabilities in a probability distribution table **must add to 1** since the table shows all of the values that the random variable can take.

For example, let X denote the score when a fair cubical die with faces marked 1, 2, 3, 4, 4 and 5 is thrown. The probability distribution of the random variable X is shown in the table below:

x	1	2	3	4	5
p	$\frac{1}{6}$	$\frac{1}{6}$	$\frac{1}{6}$	$\frac{2}{6}$	$\frac{1}{6}$

EXAMPLE 1

Cards are drawn at random and without replacement from a pack consisting of 1, 2, 2, 3, 3, 3 until a 3 is obtained. Let X denote the number of cards that must be taken. Find the probability distribution of X.

SOLUTION

X can take the values 1, 2, 3 or 4.

$P(X = 1) = P(\text{three}) = \frac{3}{6} = \frac{1}{2}$
$P(X = 2) = P(\text{not a three, three}) = \frac{3}{6} \times \frac{3}{5} = \frac{3}{10}$
$P(X = 3) = P(\text{not a three, not a three, three}) = \frac{3}{6} \times \frac{2}{5} \times \frac{3}{4} = \frac{3}{20}$
$P(X = 4) = P(\text{not, not, not, three}) = \frac{3}{6} \times \frac{2}{5} \times \frac{1}{4} \times \frac{3}{3} = \frac{1}{20}$

so the probability distribution of X is

x	1	2	3	4
p	$\frac{1}{2}$	$\frac{3}{10}$	$\frac{3}{20}$	$\frac{1}{20}$

Remember to check that the probabilities add to 1.

EXAMPLE 2

The random variable Y takes values 0, 1, 2 and 3 and is such that

$$P(Y = r) = k(r^2 + 1) \qquad \text{for } r = 0, 1, 2, 3$$

where k is a positive constant.

a) Find the value of k and hence write down a probability distribution table for Y.
b) Find $P(Y > 1)$.

SOLUTION

a) Using the rule $P(Y = r) = k(r^2 + 1)$ gives

$P(Y = 0) = k(0^2 + 1) = k$
$P(Y = 1) = k(2^2 + 1) = 2k$
$P(Y = 2) = k(2^2 + 1) = 5k$
$P(Y = 3) = k(3^2 + 1) = 10k$

The probability distribution table is

x	0	1	2	3
p	k	$2k$	$5k$	$10k$

Using the fact that the probabilities must add to 1 gives

$k + 2k + 5k + 10k = 1$
$\Rightarrow \quad 18k = 1$
$\Rightarrow \quad k = \frac{1}{18}$

You can now write down the final version of the probability distribution:

x	0	1	2	3
p	$\frac{1}{18}$	$\frac{2}{18}$	$\frac{5}{18}$	$\frac{10}{18}$

b) $P(Y > 1) = P(Y = 2 \text{ or } 3) = \frac{15}{18} = \frac{5}{6}$

Two random variables, X and Y, are said to be independent if the value taken by Y is not affected by the value taken by X.

This can be expressed formally by writing

$$P(Y = b \mid X = a) = P(Y = b)$$

or, equivalently,

$$P(X = a \text{ and } Y = b) = P(X = a) \times P(Y = b)$$

EXAMPLE 3

The random variable X has probability distribution

x	0	1	2
p	0.5	0.3	0.2

and the random variable Y has probability distribution

x	0	1	2
p	0.2	0.7	0.1

If the two random variables are independent, calculate

a) $P(X = Y)$
b) $P(X + Y = 2)$

SOLUTION

a) $P(X = Y) = P(X = 0 \text{ and } Y = 0 \text{ or } X = 1 \text{ and } Y = 1 \text{ or } X = 2 \text{ and } Y = 2)$
$= 0.5 \times 0.2 + 0.3 \times 0.7 + 0.2 \times 0.1$
$= 0.33$

b) $P(X + Y = 2) = P(X = 0 \text{ and } Y = 2 \text{ or } X = 1 \text{ and } Y = 1 \text{ or } X = 2 \text{ and } Y = 0)$
$= 0.5 \times 0.1 + 0.3 \times 0.7 + 0.2 \times 0.2$
$= 0.30$

EXERCISE 1

1 Michael and Nick decide to play a "best of five sets" match of tennis. The match finishes as soon as one player has won five sets. Michael has probability 0.6 of winning each set played and the result of each set is independent of the results of earlier sets. Let X denote the number of sets that Michael and Nick will play in their match.

a) Prove that $P(X = 3) = 0.28$.
b) Complete the probability distribution table for X:

x	3	4	5
p	0.28		

2 The random variable W takes values 2, 3, 4 and 5 and is such that

$P(W = r) = k(2r + 1)$ for $r = 2, 3, 4, 5$

a) Find the value of k and hence write down a probability distribution table for W.
b) Find $P(W < 4)$.

3 The table shows the probability distribution for a random variable Y:

x	3	4	5	6	7	8
p	0.13	0.31	0.15	0.05	q	$2q$

a) Find the value of q.
b) Find $P(Y > 5)$.
c) Find $P(4 \leqslant Y < 7)$.
d) What is the mode of this probability distribution?

4 A basket contains 6 red balls and 3 blue balls. Three of the balls are taken out of the basket at random and without replacement. Let R denote the number of red balls taken.
a) Prove that $P(R = 2) = \frac{15}{28}$.
b) Obtain the probability distribution of R.
c) Find $P(0 < R \leqslant 2)$.

5 Two fair cubical dice, each with faces marked 1, 1, 1, 2, 2 and 4, are thrown and the two scores are multiplied together to obtain a number N.
a) Prove that $P(N = 4) = \frac{5}{18}$.
b) Obtain the probability distribution of N.
c) Find the mode of this probability distribution.
d) Find $P(1 < N \leqslant 8)$.

6 A small pack of cards contains 6 red cards and 4 blue cards. Sheila chooses three of the cards at random and gives them to John. Let J denote the number of red cards given to John.
a) Obtain the probability distribution for J.
b) Find $P(J < 2)$.

7 The independent random variables U and V have probability distributions:

u	1	2	3
P(U = u)	0.2	0.2	0.6

u	1	2	3	4
P(V = u)	0.2	0.1	0.1	0.6

Calculate
a) $P(U = V)$.
b) $P(V - U = 1)$.
c) $P(UV = 4)$.

Mean, Variance and Standard Deviation of a Probability Distribution

We have already seen how a probability distribution plays the same role for a random variable as a frequency distribution plays for statistical data. We will now see how to calculate measures of central tendency and spread for probability distributions.

EXAMPLE 4

A simple fairground game consists of throwing two dice. If the total score is 9 or 10 then 20p is won and if the total score is 11 or 12 then 50p is won. Let X denote the winnings on one play of the game. Find the probability distribution of X.

The stall owner wishes to charge an entry fee that will ensure he makes an overall profit if the game is played lots of times. How much must he charge?

SOLUTION

Green die \ Red die	1	2	3	4	5	6
6	o	o	20	20	50	50
5	o	o	o	20	20	50
4	o	o	o	o	20	20
3	o	o	o	o	o	20
2	o	o	o	o	o	o
1	o	o	o	o	o	o

From the diagram it can be seen that the probability distribution of X is

x	0	20	50
p	26/36	7/36	3/36

If the game was played 3600 times you would expect results something like

x	0	20	50
f	2600	700	300

These are the "theoretical" or "expected" frequencies obtained by multiplying the probabilities by 3600. It is unlikely that you would get precisely these results if you were to play the game 3600 times, but you would certainly anticipate obtaining results that are similar to these expected frequencies.

The total expected winnings in these 3600 games would be

$$0 \times 2600 + 20 \times 700 + 50 \times 300 = 29\,000 \text{ pence}$$

You would expect the average (mean) winnings per game to be $\frac{29\,000}{3600} = 8.06$p, correct to three significant figures.

The stallholder would need to charge at least 9p per game to make a profit in the long term.

The calculation of the expected (or mean) winnings can be looked at in a different way:

$$8.06 = \frac{29\,000}{3600} = \frac{0 \times 2600 + 20 \times 700 + 50 \times 300}{3600}$$

$$= 0 \times \frac{2600}{3600} + 20 \times \frac{700}{3600} + 50 \times \frac{300}{3600}$$

$$= 0 \times \frac{26}{36} + 20 \times \frac{7}{36} + 50 \times \frac{3}{36} = \sum xp$$

This motivates the following definition:

Definition 1
For a discrete random variable X the expected value of X, or the mean value of X, is E[X] or μ, where

$$\mu = \text{E[X]} = \Sigma\, xp$$

Notice that this is very similar to the definition of the mean in descriptive Statistics:

$$\bar{x} = \frac{\Sigma\, xf}{N} \quad \text{where} \quad N = \Sigma f$$

The probability definition doesn't need a division because $\Sigma\, p = 1$.

Carrying this comparison one stage further we gain a definition for measures of spread for probability distributions. In descriptive Statistics we have:

$$\text{variance} = \frac{\Sigma\,(x-\bar{x})^2 f}{N} = \frac{\Sigma\, x^2 f}{N} - \bar{x}^2 \text{ and standard deviation} = \sqrt{\text{variance}}$$

Similarly, for probability distributions we have

Definition 2
For a random variable X the variance, Var[X] or σ^2, is given by

$$\sigma^2 = \text{Var[X]} = \Sigma\,(x-\mu)^2 p = \Sigma\, x^2 p - \mu^2$$

and the standard deviation, σ, is given by

$$\sigma = \sqrt{\text{Var[X]}} = \sqrt{\Sigma\,(x-\mu)^2 p} = \sqrt{\Sigma\, x^2 p - \mu^2}$$

Make sure you understand the shorthand: σ^2 is the abbreviation for the **variance**; σ is the abbreviation for the **standard deviation.**

EXAMPLE 5

Cards are taken at random without replacement from a pack of six cards containing cards marked 1, 2, 2, 3, 3, 3 until a 3 is obtained. Let X be the number of cards that must be taken. Find the mean and standard deviation of X.

SOLUTION

You must first find the probability distribution of X. This has already been done earlier in the chapter. Example 1 showed that the probability distribution for X is

x	1	2	3	4
p	0.5	0.3	0.15	0.05

$$\mu = \text{E[X]} = \Sigma\, xp = 1 \times 0.5 + 2 \times 0.3 + 3 \times 0.15 + 4 \times 0.05 = 1.75$$

$$\text{Var[X]} = \Sigma\, x^2 p - \mu^2 = 1^2 \times 0.5 + 2^2 \times 0.3 + 3^2 \times 0.15 + 4^2 \times 0.05 - 1.75^2 = 0.7875$$

$$\text{s.d.} = \sqrt{\text{Var[X]}} = \sqrt{0.7875} = 0.887 \text{ (3 s.f.)}$$

Your graphical or scientific calculator will probably do these calculations: input the probability distribution in the same way as you would enter a frequency distribution.

EXERCISE 2

1 The random variable X takes values 2, 4 and 12 with probabilities 1/2, 1/3 and 1/6, respectively. Find the expected value and standard deviation of X.

2 The random variable X has probability distribution

x	0	1	2	3	4
p	0.1	0.2	0.3	q	0.1

i) Find the value of q.
ii) Find the mean and standard deviation of X.

3 The probability of C winning a game against D is 1/2. The game cannot be drawn. They play a series of games until EITHER one man wins two games in succession OR one man has won a total of three games. Let X denote the number of games that must be played.
a) Prove that P(X = 3) = 0.25.
b) Prove that P(X = 5) = 0.125.

The table below shows the probability distribution of X:

x	2	3	4	5
p	0.5	0.25	0.125	0.125

c) Calculate E[X].
d) Find the standard deviation of X.

4 A basket contains six red balls and four green balls. All the balls are the same size. Pauline picks four balls at random and without replacement from the box. If W denotes the number of red balls she obtains:
a) prove that P(W = 1) = $\frac{24}{210}$.

The probability distribution of W is

w	0	1	2	3	4
p	$\frac{1}{210}$	$\frac{24}{210}$	$\frac{90}{210}$	$\frac{80}{210}$	$\frac{15}{210}$

b) Calculate the mean and variance of W.
c) Find P(W > E[W]).

5 A player throws a fair cubical die whose faces are numbered 1 to 6 inclusive. If the player obtains a 6 he throws the die a second time, and in this case his score is the sum of 6 and the second number: otherwise his score is the number obtained. The player has no more than two throws.
Let X be the random variable denoting the player's score. Write down the probability distribution of X and determine its mean and standard deviation.

6 The probability distribution of a random variable X is shown in the table below:

x	−1	0	1	2	4
p	0.1	s	t	0.2	0.2

If E[X] = 1.4:
a) determine the values of s and t;
b) find the standard deviation of X.

7 Tomorrow Fred starts three days of holiday and he wonders what sort of weather to expect. He knows that, if it is fine one day then there is a 4/5 chance of it being fine the next day, and if it is wet one day then there is a 3/5 chance of it being wet the next day. Today it is fine.
Let Y denote the number of fine days that Fred gets on holiday.
a) Prove that P(Y = 2) = 0.256.

The probability distribution for Y is shown in the table below:

y	0	1	2	3
p	0.072	0.16	0.256	0.512

b) Calculate the mean and variance of Y.

8 Lydia and Stephanie play a series of games of pool. For each game, Lydia wins with probability 0.6 and Stephanie wins with probability 0.4. The result of each game is independent of the results of the previous games. They agree to continue playing until someone has won three games. Let X denote the number of games they must play.
a) Prove that P(X = 4) = 0.3744.

The probability distribution of X is shown in the table below and has mean value μ and variance σ^2:

x	3	4	5
p	0.28	0.3744	0.3456

b) Calculate the values of μ and σ^2.

9 The probability distribution of a random variable X is shown in the table below:

u	0	1	2	4	8	16
P(U = u)	0.2	0.25	0.2	0.15	0.15	p

a) Determine the value of p.

The mean and standard deviation of the distribution are μ and σ, respectively.
b) Calculate the values of μ and σ.
c) Find $P(X > \mu + \sigma)$.

10 The random variable W has probability distribution

$$P(W = w) = k(w^2 + 1) \qquad \text{for } w = -1, 0, 1, 2$$

where k is a positive constant.
a) Prove that $k = \frac{1}{10}$.
b) Calculate the mean, μ, and the standard deviation, σ, of W.
c) Find $P(W \leqslant \mu)$.

Having studied this chapter you should know how to

- interpret a probability distribution given either as a table or as a rule and realise that $\Sigma p = 1$
- use the formulae

  $\mu = \text{E}[X] = \Sigma xp$

  $\text{Var}[X] = \Sigma x^2 p - \text{E}[X]^2$

  Standard deviation $= \sqrt{\text{Var}[X]}$

  to evaluate the mean, variance and standard deviation of a random variable X.

REVISION EXERCISE

1 The random variable X has probability distribution

x	0	1	2	3	4
p	0.2	r	0.3	0.1	s

and E[X] = 2.2.

a) Find the values of r and s.

b) Find the standard deviation of X.

2 In a game the player tosses a fair coin and throws a fair cubical die. His score is the number on the die if the coin shows tails but is twice this number if the coin shows heads.

a) Find the mean and standard deviation of the player's score.

b) Find the probability that the player's score is more than one standard deviation away from the expected value.

3 Ollie has a bag containing 3 red balls and 2 yellow balls. He takes balls out of the bag, at random and without replacement, until he has two red balls. Let J be the number of balls that he must take out of the bag.

a) Prove that P(J = 3) = 0.4.

The probability distribution for J is

j	2	3	4
p	0.3	0.4	0.3

b) Calculate the mean and standard deviation of J.

4 A discrete random variable R has the probability distribution

$$\text{P}(R = r) = k(r + 1) \qquad \text{for } r = 0, 1, 2, 3, 4$$
$$= 0 \text{ otherwise}$$

a) Find the value of the constant k.

b) Find E[R] and Var[R].

c) What is the probability that R exceeds its expected value?

5 The random variable V has probability distribution

v	1	2	3	5
p	k^2	k	$3k$	$4k^2$

a) Find the value of the constant k.
b) Calculate E[V].
c) Calculate the variance of V.

6 A bag contains seven sweets identical in shape and size. Three of the sweets are lemon sweets and four are orange sweets. Ahmar keeps selecting sweets from the bag until he gets a lemon sweet. He does not return to the bag any orange sweets which he selects. The number of sweets selected up to and including the first lemon sweet is denoted by the random variable X.

i) Show that

a) $P(X = 2) = \frac{2}{7}$.
b) $P(X = 5) = \frac{1}{35}$.

ii) The distribution of X is shown in the following table:

x	1	2	3	4	5
$P(X = x)$	$\frac{3}{7}$	$\frac{2}{7}$	$\frac{6}{35}$	$\frac{3}{35}$	$\frac{1}{35}$

Calculate

a) E[X].
b) Var[X].

(OCR Jan 2001 S1)

7 A discrete random variable X has probability distribution given in the following table. It is given that $E[X] = 0.95$:

x	−1	0	1	2	3
$P(X = x)$	a	b	0.1	0.3	0.2

i) Find the probability that X is greater than E[X].
ii) Find the values of a and b.
iii) Calculate Var[X].

(OCR Jan 2002 S1)

8 A small bag contains six tiles. Each tile has a letter written on it. Four of the tiles have the letter A written on them and two of the tiles have the letter B written on them. Nicholas removes the tiles one at a time, at random, from the bag until he has two tiles with the letter A.

Let X denote the number of tiles that Nicholas must remove from the bag. Calculate the mean and standard deviation of X.

9 Two jars containing some discs are placed on a table. Jar A contains 5 red discs and 3 white discs, and Jar B contains 2 red discs and 1 white disc. A disc is selected from Jar A at random and placed in Jar B. Two discs are then selected at random from Jar B. The first of these two discs is not replaced in Jar B before the second disc is selected. The possible outcomes are represented by the tree diagram below:

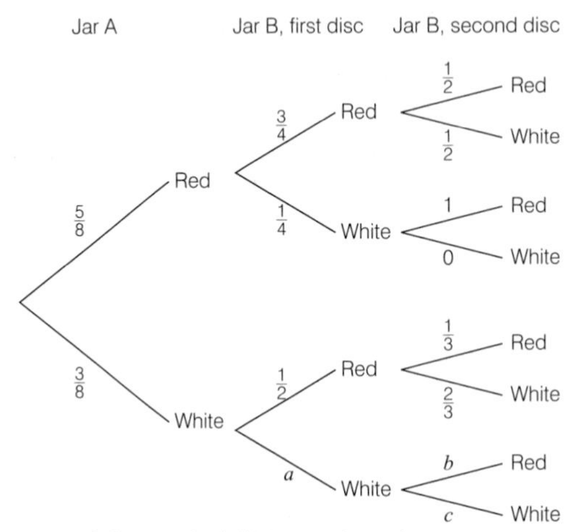

i) State the values of the probabilities a, b and c.

ii) Let D be the total number of red discs selected from Jar B. Show that $P(D = 2) = \frac{3}{8}$.

iii) The probability distribution of D is given in the table below:

d	0	1	2
$P(D = d)$	$\frac{1}{16}$	$\frac{9}{16}$	$\frac{3}{8}$

Find Var(D).

(OCR Jun 2001 S1)

10 A discrete random variable X has probability distribution given by the table below. The expectation of this distribution is denoted by μ and the variance is denoted by σ^2:

x	2	3	4	5	6
$P(X = x)$	0.2	0.1	0.1	0.4	p

i) Show that the value of p is 0.2.

ii) Find μ.

iii) Find σ^2.

(OCR Jun 2002 S1)

8 The Binomial Distribution

The purpose of this chapter is to enable you to

- calculate binomial probabilities
- appreciate when the binomial distribution is a suitable distribution for a random variable
- calculate the mean and variance of a random variable which is binomially distributed

The Probability of *r* Successes in a Sequence of Independent Experiments

Many random variables have similar probability distributions. In this module you will meet two standard probability distributions: the binomial and geometric distributions. The binomial distribution gives the probability distributions for random variables such as the number of heads when a fair coin is tossed 25 times or the number of sixes when 20 fair cubical dice are thrown.

EXAMPLE 1

Suppose that when a drawing pin is thrown upwards it has probability 0.3 of landing with the point up. Five drawing pins are thrown upwards. Let X denote the number that land with their point up. Find the probability distribution of X.

SOLUTION

Clearly X can take the values 0, 1, 2, 3, 4 or 5

Where D_1 means first pin lands point down and U_1 means first pin lands point up. Similarly D_2, U_2, etc.

$$P(X = 0) = P(D_1\ D_2\ D_3\ D_4\ D_5) = 0.7^5 = 0.16807$$

There are 5 ways in which one pin can land point up: $U_1\ D_2\ D_3\ D_4\ D_5$ or $D_1\ U_2\ D_3\ D_4\ D_5$ or $D_1\ D_2\ U_3\ D_4\ D_5$ or $D_1\ D_2\ D_3\ U_4\ D_5$ or $D_1\ D_2\ D_3\ D_4\ U_5$. Each has the same probability.

$$P(X = 1) = 5 \times P(U_1\ D_2\ D_3\ D_4\ D_5) = 5 \times 0.3 \times 0.7^4 = 0.36015$$

$$P(X = 2) = k \times P(U_1\ U_2\ D_3\ D_4\ D_5) = k \times 0.3^2 \times 0.7^3$$

where k = number of ways of arranging two Us and three Ds.

The number of ways of arranging two Us and three Ds is the same as the number of ways of picking two slots from five to put a U into and then filling the remaining slots with Ds:

EXAMPLE 1 (continued)

You can pick 2 slots from 5 in ${}_5C_2$ or 10 different ways.

So

$$P(X = 2) = {}_5C_2 \times P(U_1\, U_2\, D_3\, D_4\, D_5) = 10 \times 0.3^2 \times 0.7^3 = 0.30870$$

Similarly,

$$P(X = 3) = {}_5C_3 \times P(U_1\, U_2\, U_3\, D_4\, D_5) = 10 \times 0.3^3 \times 0.7^2 = 0.13230$$
$$P(X = 4) = {}_5C_4 \times P(U_1\, U_2\, U_3\, U_4\, D_5) = 5 \times 0.3^4 \times 0.7 = 0.02835$$
$$P(X = 5) = P(U_1\, U_2\, U_3\, U_4\, U_5) = 0.3^5 = 0.00243$$

so the probability distribution is

x	0	1	2	3	4	5
p	0.16807	0.36015	0.30870	0.13230	0.02835	0.00243

EXAMPLE 2

Suppose 12 of these drawing pins are thrown in the air and Y denotes the number that land point up. Find the probability distribution of Y.

SOLUTION

Clearly Y can take values 0, 1, 2, ..., 11, 12. It will be rather tedious to list all the probabilities and it is more practical to give a RULE:

$$\begin{aligned} P(Y = r) &= P(r \text{ UP and } 12 - r \text{ DOWN}) \\ &= {}_{12}C_r \times P(\text{first } r \text{ land UP and last } 12 - r \text{ land DOWN}) \\ &= {}_{12}C_r \times 0.3^r \times 0.7^{12-r} \end{aligned}$$

The last two examples have produced very similar probability distributions and both distributions are examples of the binomial distribution.

Definition
A random variable X is said to be binomially distributed with parameters n, p where n is a positive integer and p is a number between 0 and 1

IF the sample space of X is $\{0, 1, 2, ..., n\}$

AND $P(X = r) = {}_nC_r\, p^r(1 - p)^{n-r} = \binom{n}{r} p^r(1 - p)^{n-r}$ for $r = 0, 1, 2, 3, ..., n$

It is said that "X is B(n, p)" as shorthand for "X is binomially distributed with parameters n and p".

Looking back at Example 1 you can say that X is B(5, 0.3) and, looking at Example 2, you can say that Y is B(12, 0.3).

Now suppose that there is a series of n **independent** experiments each with two possible outcomes "SUCCESS" and "FAILURE" and each experiment has constant probability p of ending as a "SUCCESS".

Let X denote the total number of successes in the series. Clearly X can take values 0, 1, 2, ..., n.

$$P(X = r) = P(r \text{ “SUCCESSES” and } n - r \text{ “FAILURES”})$$
$$= {}_nC_r \times P(\text{first } r \text{ experiments are “SUCCESSES” and last } n - r \text{ are “FAILURES”})$$
$$= {}_nC_r \times p^r \times (1 - p)^{n-r}$$

so X is B(n, p).

Application of the Binomial Distribution

IF X = total number of successes
in a series of n **INDEPENDENT** experiments
each with two possible outcomes “SUCCESS” and “FAILURE”
and each experiment has constant probability p of resulting in a “SUCCESS”

THEN X is B(n, p)

So, for example,

If X is the number of heads when 15 fair coins are tossed then X is B(15, 0.5).
If Y is the number of sixes when 10 fair die are thrown then Y is B(10, $\frac{1}{6}$).

Experimental Verification of the Use of the Binomial Distribution

The appropriateness of a probability model can be tested by comparing experimental frequencies with the theoretical frequencies predicted by the probability model.

For example, to check whether the B(10, $\frac{1}{6}$) distribution is an appropriate probability model for the number of sixes when 10 fair dice are thrown, the experiment was performed 500 times and experimental frequencies shown in the table below obtained:

Number of sixes	0	1	2	3	4	5	6	7	8	9	10
Frequency	84	163	146	73	25	8	1	0	0	0	0

The theoretical probabilities can be calculated using the B(10, 1/6) distribution:

$$P(Y = r) = {}_{10}C_r(\tfrac{1}{6})^r(\tfrac{5}{6})^{10-r}$$

and the theoretical frequency of obtaining r sixes will then be given by

theoretical frequency = 500 × P(Y = r)

These are shown in the table below:

Number of sixes, r	0	1	2	3	4	5	6	7	8	9	10
P(Y = r)	0.162	0.323	0.291	0.155	0.054	0.013	0.002	0.000	0.000	0.000	0.000
Theoretical frequency	80.8	161.5	145.4	77.5	27.1	6.5	1.1	0.1	0.0	0.0	0.0

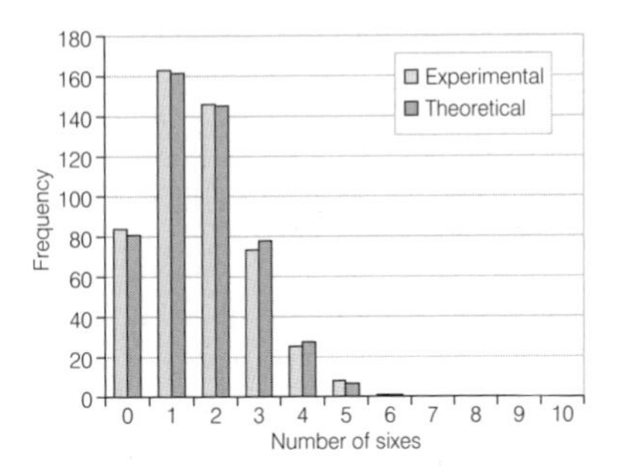

The tables and the graph show that the theoretical frequencies predicted by the B(10, 1/6) probability distribution are in good agreement with the frequencies obtained experimentally, so the B(10, 1/6) distribution is a good model for the number of sixes when 10 fair dice are thrown.

Calculations with the Binomial Distribution

EXAMPLE 3

A biased coin has probability 0.64 of landing on heads. If this coin is tossed five times find the probability of getting at least four heads.

SOLUTION

Let X = number of heads obtained. X is B(5, 0.64).

P(at least 4 heads) = P(X = 4 or 5)

$= {}_5C_4 \times 0.64^4 \times 0.36 + {}_5C_5 \times 0.64^5 \times 0.36^0$

$= 0.30198 \ldots + 0.10737 \ldots$

$= 0.409$ (3 s.f.)

EXAMPLE 4

Jean knows that if she throws a dart at a dart board she has probability 0.12 of the dart landing in the "20" region. She decides to throw five darts at the board. Let Y be the number of darts that land in the "20" region. Y can be modelled by the B(n, p) distribution.

a) State the values of the parameters n and p. Comment on the assumptions that must be made in order for this to be an appropriate model.

b) Use this model to estimate the probability that two of Jean's darts land in the "20" region.

SOLUTION

a) Y can be modelled by B(5, 0.12) provided:

- the landing place of each dart is independent of the landing points of the other darts;
- each dart has the same probability of landing in the "20" region.

If she throws the second and later darts without removing the earlier darts, it is unlikely that either of the above conditions will be strictly true: one dart being in the "20" region may well affect the likelihood of the next dart being there. It may be that her accuracy may increase with success or it may be that a dart being in the "20" region may block another dart from landing there. You should certainly therefore assume she removes one dart before throwing the next.

When you discuss the assumptions being made in applying a particular probability model **it is important your comments take into account the context of the situation**. In this case it would not be enough just to say that independence and constant probability are being assumed.

b) $P(Y = 2) = {}_5C_2 \times 0.12^2 \times 0.88^3 = 0.0981$ (3 s.f.)

EXAMPLE 5

Packets of Christmas tree lights contain 6 lights. Each light has probability p of being defective. Long-term tests have shown that the probability of a packet containing two defective lights is one-third of the probability of a packet containing just one defective light.

a) Find the value of p.

John has a Christmas tree and wishes to have 15 lights on it.

b) What is the probability that he will be able to fully light the tree if he buys three packets of the lights?

What assumptions have been made in answering this question?

SOLUTION

a) Let X = number of defective lights in a packet. X is B(6, p).

You are told

$$
\begin{aligned}
& P(X=2) = \tfrac{1}{3} \times P(X=1) \\
\Rightarrow\quad & 3 \times P(X=2) = P(X=1) \\
\Rightarrow\quad & 3 \times {}_6C_2\, p^2(1-p)^4 = {}_6C_1\, p^1(1-p)^5 \\
\Rightarrow\quad & 45\, p^2(1-p)^4 = 6p(1-p)^5 \\
\Rightarrow\quad & 45p = 6(1-p) \\
\Rightarrow\quad & 51p = 6 \\
\Rightarrow\quad & p = \tfrac{2}{17}
\end{aligned}
$$

b) Let Y = number of defective lights in three packets. Y is B(18, $\frac{2}{17}$).
If the tree is to be fully lit John must have at least 15 good lights – so he must have at most 3 defective lights in the three packets.

$$
\begin{aligned}
P(Y \leqslant 3) &= P(Y = 0, 1, 2, 3) \\
&= {}_{18}C_0 \times (\tfrac{2}{17})^0 \times (\tfrac{15}{17})^{18} + {}_{18}C_1 \times (\tfrac{2}{17})^1 \times (\tfrac{15}{17})^{17} \\
&\quad + {}_{18}C_2 \times (\tfrac{2}{17})^2 \times (\tfrac{15}{17})^{16} + {}_{18}C_3 \times (\tfrac{2}{17})^3 \times (\tfrac{15}{17})^{15} \\
&= 0.10509\ldots + 0.25221\ldots + 0.28584\ldots + 0.20326\ldots \\
&= 0.846
\end{aligned}
$$

For the binomial distribution to be an appropriate model you need to be sure that:

- each bulb has the **same** probability of being defective;
- one bulb being defective is independent of any other bulb in the packet being defective.

Deciding on whether these are realistic assumptions would probably depend upon knowledge about the production and packing processes.

EXERCISE 1

1 A fair coin is tossed 9 times. Find the probability of obtaining exactly three Heads.

2 A large town contains 55% males and 45% females. If a sample of seven people is chosen at random from the town, find the probability that just two of the sample are female.

3 A multiple choice test consists of 20 questions each with five possible answers. Andy has not done any revision for the test so guesses the answer for each question. Find the probability that he answers five or six questions correctly.

4 Jack has 15 fair cubical dice. Each die has the numbers 1, 2, 3, 4, 5 and 6 marked on the faces. If Jack throws the 15 dice together, find the probability that he will obtain three or four 6s.

5 28% of the adults living in Britain are regular smokers. Estimate the probability that a party of 20 British adults will contain 4 or 5 regular smokers.
Comment critically on the assumptions that have been made in answering this question.

6 Michael catches the bus to work each day. Over a long period he has recorded that 32% of the buses arrive at the bus-stop near his house at least five minutes later than they are supposed to. Let Y be the number of times during 20 working days that Michael's bus is at least five minutes later than it is supposed to be.

Suggest a suitable probability model for Y and state two assumptions which are required to make this a good model.

Estimate the probability that during the next period of 4 weeks (i.e. 20 working days) his bus will be at least five minutes late arriving at the bus-stop on nine or ten occasions.

7 A biased coin has probability p of landing on heads, where $0 < p < 1$. The coin is more likely to land on heads than it is to land on tails.

a) Write down an expression for the probability of the coin landing on heads three times out of four tosses.

b) Write down an expression for the probability of the coin landing on heads four times out of six tosses.

The probability that the coin lands on heads three times out of four tosses is twice the probability that the coin lands on heads four times out of six tosses.

c) Show that p must satisfy the equation $15p^2 - 15p + 2 = 0$ and hence find the value of p correct to four decimal places.

d) If this coin is tossed 10 times find the probability that it will land on heads on exactly eight occasions.

8 The random variable W is B(15,0.72). Calculate the probability that the value of W is 0, 6 or 12.

Efficient Calculation of Binomial Probabilities

Use of Binomial Tables

Your A level formula booklet gives tables of CUMULATIVE binomial probabilities for binomial distributions with a wide range of values of n and p. These tables can reduce significantly the amount of work involved in evaluating binomial probabilities.

The table below shows an extract of these tables in the case when $n = 10$:

$n = 10$ p		0.05	0.1	0.15	1/6	**0.2**	0.25	0.3	1/3	0.35	0.4
$x = 0$	0	0.5987	0.3487	0.1969	0.1615	**0.1074**	0.0563	0.0282	0.0173	0.0135	0.0060
1	1	0.9139	0.7361	0.5443	0.4845	**0.3758**	0.2440	0.1493	0.1040	0.0860	0.0464
2	2	0.9885	0.9298	0.8202	0.7752	**0.6778**	0.5256	0.3828	0.2991	0.2616	0.1673
3	3	0.9990	0.9872	0.9500	0.9303	**0.8791**	0.7759	0.6496	0.5593	0.5138	0.3823
4	4	0.9999	0.9984	0.9901	0.9845	**0.9672**	0.9219	0.8497	0.7869	0.7515	0.6331
5	5	1.0000	0.9999	0.9986	0.9976	**0.9936**	0.9803	0.9527	0.9234	0.9051	0.8338
6	6	1.0000	1.0000	0.9999	0.9997	**0.9991**	0.9965	0.9894	0.9803	0.9740	0.9452
7	7	1.0000	1.0000	1.0000	1.0000	**0.9999**	0.9996	0.9984	0.9966	0.9952	0.9877
8	8	1.0000	1.0000	1.0000	1.0000	**1.0000**	1.0000	0.9999	0.9996	0.9995	0.9983
9	9	1.0000	1.0000	1.0000	1.0000	**1.0000**	1.0000	1.0000	1.0000	1.0000	0.9999
10	10	1.0000	1.0000	1.0000	1.0000	**1.0000**	1.0000	1.0000	1.0000	1.0000	1.0000

The shaded column gives the cumulative probabilities for a random variable X which is B(10, 0.2).

From the table you can see that:

$P(X \leqslant 0)$	0.1074
$P(X \leqslant 1)$	0.3758
$P(X \leqslant 2)$	0.6778
$P(X \leqslant 3)$	0.8791
$P(X \leqslant 4)$	0.9672
$P(X \leqslant 5)$	0.9936
$P(X \leqslant 6)$	0.9991
$P(X \leqslant 7)$	0.9999
$P(X \leqslant 8)$	1.0000
$P(X \leqslant 9)$	1.0000
$P(X \leqslant 10)$	1.0000

and you can easily deduce other probabilities for X:

$$P(X = 5) = P(X \leqslant 5) - P(X \leqslant 4) = 0.9936 - 0.9672 = 0.0264$$

$$\begin{aligned} P(X \geqslant 2) &= P(X = 2, 3, 4, 5, 6, 7, 8, 9, 10) \\ &= 1 - P(X = 0, 1) \\ &= 1 - P(X \leqslant 1) \\ &= 1 - 0.3758 = 0.6242 \end{aligned}$$

$$\begin{aligned} P(3 \leqslant X \leqslant 6) &= P(X = 3, 4, 5, 6) \\ &= P(X \leqslant 6) - P(X \leqslant 2) \\ &= 0.9991 - 0.6778 \\ &= 0.3213 \end{aligned}$$

$$\begin{aligned} P(2 < X \leqslant 5) &= P(X = 3, 4, 5) \\ &= P(X \leqslant 5) - P(X \leqslant 2) \\ &= 0.9936 - 0.6778 \\ &= 0.3158 \end{aligned}$$

EXAMPLE 6

35% of the electorate in a large town are Conservative supporters. If a random sample of 20 people from the town is taken, find the probability that the sample contains:

a) at most 9 Conservatives;
b) exactly 8 Conservatives;
c) more than 5 Conservatives.

Find also the largest value k for which the statement

"there is a greater than 95% chance that the sample contains at least k Conservative supporters"

is true.

EXAMPLE 6 (continued)

SOLUTION

Let Y = number of Conservatives in the sample. Y is B(20, 0.35)

a) $P(Y \leqslant 9) = 0.8782$

b) $P(Y = 8) = P(Y \leqslant 8) - P(Y \leqslant 7)$
$= 0.7624 - 0.6010 = 0.1614$

c) $P(Y > 5) = 1 - P(Y = 0, 1, 2, 3, 4, 5)$
$= 1 - P(Y \leqslant 5) = 1 - 0.2454 = 0.7546$

For the final part, you want the largest value of k such that $P(Y \geqslant k) > 0.95$.

Reading down the cumulative probability table you can see that

$$P(Y = 0) = 0.0002 \quad \Rightarrow \quad P(Y \geqslant 1) = 0.9998$$
$$P(Y \leqslant 1) = 0.0021 \quad \Rightarrow \quad P(Y \geqslant 2) = 0.9979$$
$$P(Y \leqslant 2) = 0.0121 \quad \Rightarrow \quad P(Y \geqslant 3) = 0.9879$$
$$P(Y \leqslant 3) = 0.0444 \quad \Rightarrow \quad P(Y \geqslant 4) = 0.9556$$
$$P(Y \leqslant 4) = 0.1182 \quad \Rightarrow \quad P(Y \geqslant 5) = 0.8818$$

So the required value of k is 4.

Extract of cumulative probabilities for a variable, X, which is B(20, 0.35)

r	$P(X \leqslant r)$
0	0.0002
1	0.0021
2	0.0121
3	0.0444
4	0.1182
5	0.2454
6	0.4166
7	0.6010
8	0.7624
9	0.8782
10	0.9468
11	0.9804
12	0.9940
13	0.9985
14	0.9997
15	1.0000
16	1.0000
17	1.0000
18	1.0000
19	1.0000
20	1.0000

Using a Graphical Calculator

If your graphical calculator can handle sigma notation then it can be used to perform lengthy binomial calculations.

For example, if you wish to find $P(Y \leqslant 9)$ where Y is B(20, 0.35) you can write

$$P(Y \leqslant 9) = P(Y = 0, 1, 2, \ldots, 9) = \sum_{r=0}^{20} {}_{20}C_r \times 0.35^r \times 0.65^{20-r} = 0.878 \qquad \text{(3 s.f.)}$$

and

$$P(Y > 5) = P(Y = 6, 7, \ldots, 20) = \sum_{r=6}^{20} {}_{20}C_r \times 0.35^r \times 0.65^{20-r} = 0.755 \qquad \text{(3 s.f.)}$$

EXAMPLE 7

A bulb manufacturer produces a large number of tulip bulbs and packs them randomly in boxes of 10. 34% of the bulbs will produce red flowers.

a) Calculate the probability that a box of tulips will contain at least three bulbs that will produce red flowers.

b) Bert Greenfingers buys 15 boxes of tulip bulbs. Find the probability that exactly nine of these boxes will contain at least three bulbs that will produce red flowers.

SOLUTION

a) Let X = number of bulbs producing red flowers in a box

X is B(10, 0.34)

$$P(X \geqslant 3) = P(X = 3, 4, \ldots, 10) = \sum_{r=3}^{10} {}_{10}C_r \times 0.34^r \times 0.66^{10-r} = 0.7162 \quad (4 \text{ d.p.})$$

If you haven't a graphical calculator, you can do this as

$$\begin{aligned} P(X \geqslant 3) &= 1 - P(X = 0, 1, 2) \\ &= 1 - ({}_{10}C_0 \times 0.34^0 \times 0.66^{10} + {}_{10}C_1 \times 0.34^1 \times 0.66^9 + {}_{10}C_2 \times 0.34^2 \times 0.66^8) \\ &= 1 - (0.01568\ldots + 0.08079\ldots + 0.18729\ldots) \\ &= 0.7162 \quad (4 \text{ d.p.}) \end{aligned}$$

b) If Y = number of boxes that Bert buys with at least three red bulbs then Y is B(15, 0.7162)

$$P(Y = 9) = {}_{15}C_9 \times 0.7162^9 \times 0.2838^6 = 0.130 \quad (3 \text{ d.p.})$$

EXAMPLE 8

How many times must a fair cubical die be thrown in order to be 99% certain of getting at least one six?

SOLUTION

Suppose you have to throw the die N times. This produces

$$\begin{aligned} & P(\text{at least 1 six}) \geqslant 0.99 \\ \Rightarrow \quad & 1 - P(\text{no sixes}) \geqslant 0.99 \\ \Rightarrow \quad & P(\text{no sixes}) \leqslant 0.01 \\ \Rightarrow \quad & (\tfrac{5}{6})^N \leqslant 0.01 \end{aligned}$$

If you have studied module C2, this inequality may be solved by taking logs:

$$\begin{aligned} \Rightarrow \quad & \log[(\tfrac{5}{6})^N] \leqslant \log 0.01 \\ \Rightarrow \quad & N \log[(\tfrac{5}{6})] \leqslant \log 0.01 \\ \Rightarrow \quad & -0.07918/N \leqslant -2 \\ \Rightarrow \quad & N \geqslant 25.3 \end{aligned}$$

Remember, dividing by a negative number reverses the direction of the inequality.

so 26 throws are necessary.

(If you haven't yet studied logarithms, you could solve the inequality $(\frac{5}{6})^N \leqslant 0.01$ using a trial and improvement method.)

EXERCISE 2

1 Find the probability of:
a) exactly three heads when a fair coin is tossed eight times;
b) exactly two sixes when a fair cubical die is rolled 15 times;
c) at least two heads when a fair coin is tossed six times.

2 When Bill and Fred play pool Bill has probability 0.6 of winning any game. No game can end as a draw. They play a series of 7 games. Find the probability that Bill wins:
a) exactly 5 games; **b)** at most 5 games.

3 Whenever I ring James there is a 0.4 chance of him being able to answer the phone. If I ring him 10 times find the probability that he answers:
a) at most 5 times; **b)** exactly 3 times; **c)** at least 3 times.

4 When Jane plants a daffodil bulb there is a probability of 0.75 that the bulb will produce a flowering plant. If Jane plants 10 bulbs, find the probability that:
a) at most 5 of the bulbs produce flowers;
b) exactly 3 of the bulbs produce flowers;
c) at least 3 of the bulbs produce flowers.

5 Jack throws a fair die 10 times. Find the probability that he gets exactly two sixes.
A group of eight boys each throw a fair die 10 times. Find the probability that exactly three of the eight boys obtain exactly two sixes.

6 **a)** A large factory produces electrical fuses, 10% of which are defective. If I take 20 fuses at random from the production line, find the probability of there being at least two defective fuses.
b) A box of 50 fuses contains five which are defective. If I take a sample of 20 out of the box, calculate the probability of there being at least two defective fuses.

Explain why your answers to (a) and (b) are different.

7 How many times must I toss a fair coin in order to be at least 99.9% certain of obtaining at least one head?

8 **a)** A random variable W has a B(4, 0.4) distribution. Calculate the probability that the value of W is an odd number.
b) The random variable X has a B(10, p) distribution where $0 < p < 1$. Given that

$$P(X = 8) = 2P(X = 9)$$

find the value of p.

9 A biased die has probability p of landing on six. If the die is thrown 10 times, the probability of getting exactly two sixes is three times the probability of getting exactly three sixes. Find the value of p.
Find the probability of two sixes in 18 throws of this die.
How many times must I throw this die to be 95% certain of obtaining at least one six?

10 A flowerbed is planted with 20 seedlings. Each seedling has a 0.7 chance of developing into a mature plant. Find the probability that:
a) exactly 16 seedlings mature;
b) at least 12 seedlings mature.

If seven flowerbeds are each planted with 20 seedlings find the probability that at least 12 seedlings mature in exactly five of the flowerbeds.

11 Large batches of similar components are delivered to a company. A sample of 5 articles is taken at random from each batch and tested to destruction. If at least 4 out of the 5 articles meet the company's requirements then the batch is accepted. Otherwise the batch is rejected.
Let p denote the probability that a component doesn't meet the company's requirements. Show that the probability that the company accepts a batch is $A(p)$ where

$$A(p) = (1 - p)^4(1 + 4p)$$

Use a graphical calculator or computer graph drawing package to produce a graph of $A(p)$ for values of p from 0 to 1. Obtain the value of p for which $A(p) = 0.95$.

The Mean and Variance of the Binomial Distribution

The mean and variance of random variables that have the binomial distribution may be calculated in the normal fashion, using the formulae

$$\mu = E[X] = \sum xp \qquad Var[X] = \sum x^2 p - \mu^2$$

EXAMPLE 9

Find the mean and variance of:

a) U which is B(4, 0.7);
b) V which is B(6, 0.2);
c) W which is B(5, 0.4).

SOLUTION

a) The table below shows the probability distribution of U:

u	0	1	2	3	4
p	0.0081	0.0756	0.2646	0.4116	0.2401

Using a calculator gives

$\mu = E[U] = \sum up = 2.8$
$Var[U] = \sum u^2 p - \mu^2 = 8.68 - 2.8^2 = 0.84$

b) The random variable V, which is B(6, 0.2), has probability distribution table

v	0	1	2	3	4	5	6
p	0.262144	0.393216	0.245760	0.081920	0.015360	0.001536	0.000064

which gives

$\mu = E[V] = \sum vp = 1.2$
$Var[V] = \sum v^2 p - \mu^2 = 2.4 - 1.2^2 = 0.96.$

c) The probability distribution table for W is

x	0	1	2	3	4	5
p	0.07776	0.25920	0.34560	0.23040	0.07680	0.01024

So

$\mu = E[X] = \sum xp = 2.00000$
$Var[X] = \sum x^2 p - E[X]^2 = 5.2 - 2^2 = 1.2$

In example 9 the mean and variance of three different binomial distributions have been calculated. Collecting these results together in a table may show an overall pattern:

n	p	μ		**Variance**
5	0.4	2	$\xrightarrow{\times 0.6}$	1.2
4	0.7	2.8	$\xrightarrow{\times 0.3}$	0.84
6	0.2	1.2	$\xrightarrow{\times 0.8}$	0.96

Notice that in each case, the mean can be obtained by multiplying n by p and the variance can be obtained by multiplying the mean by $(1 - p)$.

Generalising these results gives

If X is $B(n, p)$ then

$$\mu = E[X] = np$$

and

$$Var[X] = np(1 - p)$$

These results should be used in future when dealing with the mean and variance of a binomial distribution. The results can be proved using ideas from the C2 and C3 modules: the proof is given in the Extension to this chapter.

EXAMPLE 10

A certain disease afflicts 3% of the population. A test for this disease gives a positive result on 95% of people who have the disease. The test gives a positive result on 2.5% of people who don't have the disease.

a) A person is taken at random from the population. Find the probability that they give a positive response to the disease.

During the course of a day the clinic tests 20 people for the disease. Assuming that this sample is a random sample taken from the whole population, find:

b) the probability of getting at least two positive responses;
c) the mean and variance of the number of positive responses.

SOLUTION

a) P(positive response) = P(has disease and positive OR no disease and positive)
$= 0.03 \times 0.95 + 0.97 \times 0.025 = 0.05275$

b) Let X = number in sample who give positive response. X is B(20, 0.05275).

$$\begin{aligned} P(X \geqslant 2) &= 1 - P(X = 0 \text{ or } 1) \\ &= 1 - (0.94725^{20} + {}_{20}C_1 \times 0.05275 \times 0.94725^{19}) \\ &= 1 - [0.33829\ldots + 0.37677\ldots] = 0.285 \quad \text{(3 s.f.)} \end{aligned}$$

c) $E[X] = np = 20 \times 0.05275 = 1.055$
$Var[X] = np(1 - p) = 20 \times 0.05275 \times 0.94725 = 0.9993 \quad \text{(4 s.f.)}$

EXAMPLE 11

The random variable Y is B(16, 0.7). If μ is the expected value of Y and σ is the standard deviation of Y, use binomial tables to evaluate $P(Y > \mu + \sigma)$.

SOLUTION

$$\mu = E[Y] = np = 16 \times 0.7 = 11.2$$
$$Var[Y] = np(1 - p) = 16 \times 0.7 \times 0.3 = 3.36$$
$$\Rightarrow \quad \sigma = \sqrt{3.36} = 1.833 \quad \text{(3 d.p.)}$$

$$\begin{aligned} P(Y > \mu + \sigma) &= P(Y > 13.033 \ldots) \\ &= P(Y \geqslant 14) \\ &= 1 - P(Y \leqslant 13) \\ &= 1 - 0.9006 \\ &= 0.0994 \end{aligned}$$

EXERCISE 3

1 Find the mean and standard deviation of:

a) the number of heads when a fair coin is tossed 40 times;

b) the number of sixes when a fair cubical die with faces 1, 2, 3, 4, 5 and 6 is thrown 90 times;

c) the number of smokers in a random sample of 200 people if 28% of the population are smokers.

2 There are four flights each day from Cardiff to Amsterdam. The probability that any flight is more than five minutes late arriving in Amsterdam is 0.12. Let W denote the number of flights in a week that are more than five minutes late arriving in Cardiff.

a) W is to be modelled by $B(n, p)$.

i) What are the values of the parameters n and p?

ii) State two assumptions, in context, that must be made for this model to be appropriate.

b) Use this model to calculate:

i) $P(W > 2)$;

ii) $E[W]$;

iii) the standard deviation of W.

3 The random variable X is $B(n, p)$ and has mean 12 and variance 2.4.

a) Calculate the values of n and p.

b) Find $P(X > 11)$.

4 Mr Jones has two wardrobes: in the first wardrobe he has four red ties and six blue ties; in the second wardrobe he has three blue ties and seven red ties. Each morning, when he gets up, Mr Jones throws a fair cubical die with faces marked 1, 2, 3, 4, 5 and 6. If the result is 1 or 2 then he picks his tie at random from wardrobe one; if the result is 3, 4, 5 or 6 then he picks his tie at random from wardrobe two.

a) Find the probability, p, that Mr Jones picks a blue tie in the morning.

Let R be the number of times that Mr Jones wears a red tie in a period of 10 days.

b) State one assumption that must be made in order that T can be modelled by $B(10, p)$.

c) Find the mean, μ, and the standard deviation, σ, of T.

d) Find $P(T < \mu - \sigma)$.

5 The random variable Y is $B(n, p)$ and has mean 16 and variance 3.2.

a) Calculate the values of n and p.

b) Find $P(13 \leqslant Y < 18)$.

6 The probability that Pauline's favourite parking space at work is occupied by another car when she arrives at work is 0.35. Let T be the number of days in four weeks (i.e. 20 working days) that Pauline arrives at work only to find that her favourite parking space is occupied.

a) State two assumptions, in context, that must be satisfied for the binomial distribution to be an appropriate model for T and state the values of the parameters.

Assuming that the binomial distribution is appropriate:

b) find the mean, μ, and the standard deviation, σ, of T;

c) find, using tables or otherwise, $P(T > \mu + 2\sigma)$.

EXTENSION

Proof of Mean and Variance of Binomial Distribution

Let X be $B(n, p)$ so that

$$P(X = r) = {}_nC_r p^r(1-p)^{n-r} \qquad \text{for } r = 0, 1, 2, 3, \ldots, n$$

Recall from the work on the binomial expansion in C2 that

$$(q+pt)^n = q^n + {}_nC_1 q^{n-1}pt + {}_nC_2 q^{n-2}p^2t^2 + {}_nC_3 q^{n-3}p^3t^3 + \cdots + {}_nC_r q^{n-r}p^rt^r + \cdots + p^nt^n$$
$$= \sum_{r=0}^{n} {}_nC_r q^{n-r}p^rt^r$$

We can now differentiate each side of this equation with respect to t. The chain rule (module C3) tells us that the derivative of $(q+pt)^n$ is $np(q+pt)^{n-1}$ so we have

$$np(q+pt)^{n-1} = \sum_{r=0}^{n} r\,{}_nC_r q^{n-r}p^rt^{r-1}$$

Now substitute $t = 1$ and $q = 1 - p$ into this formula to obtain

$$np(1-p+p)^{n-1} = \sum_{r=0}^{n} r\,{}_nC_r p^r(1-p)^{n-r}$$

$$\Rightarrow \quad np = \sum_{r=0}^{n} r\,{}_nC_r p^r(1-p)^{n-r} = \sum_{r=0}^{n} rP(X = r) = E[X]$$

so we have proved that $\mu = E[X] = np$.

Returning to the formula

$$np(q+pt)^{n-1} = \sum_{r=0}^{n} r\,{}_nC_r q^{n-r}p^rt^{r-1}$$

multiplying through by t gives

$$npt(q+pt)^{n-1} = \sum_{r=0}^{n} r\,{}_nC_r q^{n-r}p^rt^r$$

and differentiating again, using the product rule and the chain rule on the left-hand side, we obtain

$$np(q+pt)^{n-1} + npt(n-1)p(q+pt)^{n-2} = \sum_{r=0}^{n} r^2 {}_nC_r q^{n-r} p^r t^{r-2}$$

Now substitute $t = 1$ and $q = 1 - p$ into this formula to obtain

$$np(1-p+p)^{n-1} + np(n-1)p(1-p+p)^{n-2} = \sum_{r=0}^{n} r^2 {}_nC_r (1-p)^{n-r} p^r$$

$$\Rightarrow \quad np + n(n-1)p^2 = \sum_{r=0}^{n} r^2 {}_nC_r p^r (1-p)^{n-r} = \sum r^2 \text{P}(\text{X} = r)$$

Now $\text{Var[X]} = \sum r^2 \text{P}(\text{X} = r) - \mu^2$

$\Rightarrow \quad \text{Var[X]} = np + n(n-1)p^2 - (np)^2$

$\Rightarrow \quad \text{Var[X]} = np + n^2p^2 - np^2 - n^2p^2$

$\Rightarrow \quad \text{Var[X]} = np - np^2$

$\Rightarrow \quad \text{Var[X]} = np(1-p)$ as required

Having studied this chapter you should know how to

- **define** the binomial distribution:
  a random variable X is said to have the binomial distribution with parameters n and p, where n is a positive integer and $0 < p < 1$
  if

  X can take values 0, 1, 2, ..., n

  and

  $\text{P}(\text{X} = r) = {}_nC_r p^r (1-p)^{n-r}$

  and if this is the case you can write "X is B(n, p)"

- **use** the binomial distribution:

  IF X = total number of successes
  in a series of n **independent** experiments
  each with two possible outcomes "SUCCESS" and "FAILURE"
  and each experiment has **constant probability** p of ending as a "SUCCESS"

  THEN X is B(n, p)

- use the formula $\text{P}(\text{X} = r) = {}_nC_r p^r (1-p)^{n-r}$, the binomial tables and your graphical calculator to evaluate binomial probabilities

- calculate the mean and variance of a random variable, X, which is B(n, p) using the formulae E[X] = np and Var[X] = $np(1-p)$

REVISION EXERCISE

1 Every day Nicholas tries to telephone his mother. Over the past 200 days he was successful on the first attempt on 130 occasions. During the next fortnight, he will try to telephone his mother each day. Let S be the number of days, out of 14, that he is successful at the first attempt.

a) Suggest a suitable model for S, giving the values of any parameters.
b) State two assumptions, in context, which are required to make this a good model.
c) Calculate $P(S = 9)$.
d) Calculate $P(S < 12)$.

2 Last month Mrs Jones entered 18 postal competitions. The probability of her winning a prize in any particular competition is 0.05, independently of the other competitions. Let Y denote the number of prizes she wins.

a) The distribution of Y is $B(n, p)$. State the values of the parameters n and p.
b) Calculate $P(Y = 2)$.
c) Use binomial tables to find:
  i) $P(Y \geqslant 2)$;
  ii) $P(Y < 4)$.

3 The random variable X has a $B(18, 0.4)$ distribution. Using the tables of cumulative binomial probabilities, and giving your answers correct to 3 significant figures, find:

i) $P(X \leqslant 7)$;
ii) $P(X > 6)$;
iii) $P(8 < X \leqslant 14)$.

(OCR Jan 2002 S1)

4 Over a period of time it has been noted that 90% of trains running from Cardiff to London arrive in London within 5 minutes of their expected arrival time. Tomorrow 16 trains will run from Cardiff to London. Let X denote the number of these trains that arrive in London within 5 minutes of their expected arrival time.

a) Suggest a suitable probability model for X, giving the values of any parameters.
b) State two assumptions, in context, which are required to make this a good model.
c) Using binomial tables or a calculator, determine:
  i) $P(X > 12)$;
  ii) $P(11 \leqslant X \leqslant 15)$.
d) What is the largest value of k satisfying the condition that $P(X \geqslant k) \geqslant 0.95$?
e) Find $P(X > E[X])$.

5 Lily has a die which has probability 0.4 of landing on a six.
Nora has a die which has probability 0.2 of landing on a six.
Lily throws her dice three times. Let L be the number of sixes she obtains.
Nora throws her dice five times. Let N be the number of sixes she obtains.
Calculate:

a) $P(L = 2)$;
b) $P(N > 1)$;
c) $P(L = N)$.

6 The random variable Y is $B(15, p)$.

a) If $E[Y] + Var[Y] = 12.6$ determine the value of p.
b) Find $P(Y > 10)$.

7 In the past soccer player Michael has been successful with 85% of his penalty kick attempts at goal. Each day next week he is going to practice taking penalty kicks and intends to take 20 practice penalties each day from Monday to Friday.

Let X be the number of successful kicks on a particular day.

a) State suitable values for n and p for X to be modelled by $B(n, p)$.

b) State two assumptions, in context, which are required to make this a good model.

c) Use binomial tables to obtain $P(X \geqslant 18)$.

Let Y be the number of days next week on which Michael is successful with at least 18 of his kicks.

d) Suggest a suitable probability model for Y, giving the values of any parameters and state one assumption that is required to make this a good model.

e) Find $P(Y = 2)$.

f) Calculate the mean and standard deviation of Y.

8 The percentage of cars on British roads that are red is $100p\%$ where $0 < p < 1$.

a) Write down an expression involving p for the probability that 5 of the next 20 cars to pass a bridge on the M1 are red.

b) Write down an expression involving p for the probability that 4 of the next 20 cars to pass a bridge on the M1 are red.

The probability that 5 of the next 20 cars to pass a bridge on the M1 are red is twice the probability that 4 of the next 20 cars to pass this bridge are red.

c) Write down and solve an equation that must be satisfied by p.

d) Find the probability that 5 of the next 10 cars that pass the bridge are red.

e) Alan, Brian, Colin, David and Edward each conduct a survey at the same time of 10 cars at 5 bridges on 5 different motorways. Find the probability that two of them see exactly 5 red cars.

9 a) The random variable W has a $B(6, 0.4)$ distribution. Calculate the probability that the value of W is an odd number.

b) The random variable X has a $B(5, p)$ distribution where $p \neq 0$. Given that $E[X] = 3\text{Var}[X]$, find $P(X = 0)$.

c) The random variable Y has a $B(12, p)$ distribution where $p \neq 0$. Given that $P(Y = 11) = P(Y = 12)$ find the value of p.

(OCR Jan 2001 S1)

10 Sheena travels to work by car. From long observation, she has found that she can park in her favourite parking space on 2 days out of 5 on average. Let X be the number of days out of a 5-day working week on which she can park in her favourite parking space.

i) State two assumptions which need to be made for a binomial model to be valid for the distribution of the random variable X.

ii) Assuming that $B(n, p)$ is a valid model for the distribution of X:

a) state the values of the parameters n and p;

b) show that $P(X > 3) = 0.0870$ correct to 3 significant figures.

iii) A 5-day working week in which Sheena can park in her favourite parking space on more than 3 days is a "good" week. Find the probability that, out of 7 randomly chosen 5-day working weeks, fewer than 2 are good weeks.

(OCR Jun 2001 S1)

9 The Geometric Distribution

The purpose of this chapter is to enable you to

- calculate geometric distribution probabilities
- appreciate when the geometric distribution is a suitable distribution for a random variable
- calculate the mean of a random variable which is geometrically distributed

The geometric distribution is the second of the standard probability distributions that will be met in this module. Many random experiments will be repeated time after time until a certain event, which might be called a success, occurs. For example, at the beginning of some board games a six must be thrown with a dice before a player can start to move his counter around the board or at a fairground a child may keep playing a game until she wins a goldfish. The geometric distribution provides an appropriate probability model for the number of goes at the random experiment that are necessary to record the first success.

Suppose that a biased die has probability 0.25 of landing on a six and that the die is thrown repeatedly until a six is obtained. Let Y denote the number of throws required up to and including the first six.

The random variable Y can take any positive integer value.

Let S_i denote the ith throw being a six and N_i denote the ith throw not being a six.

$$P(Y = 1) = P(S_1) = 0.25$$
$$P(Y = 2) = P(N_1 S_2) = 0.75 \times 0.25$$
$$P(Y = 3) = P(N_1 N_2 S_3) = 0.75 \times 0.75 \times 0.25 = 0.75^2 \times 0.25$$

We can multiply the probabilities here since the result of each throw is independent of the results of the previous throws,

and in general

$$P(Y = r) = P(N_1 N_2 \ldots N_{r-1} S_r) = 0.75^{r-1} \times 0.25$$

so the probability distribution for Y can be summarised by the rule

$$P(Y = r) = 0.75^{r-1} \times 0.25 \qquad r = 1, 2, 3, \ldots$$

This is an example of a distribution known as the **geometric distribution**

Definition

A random variable X is said to have the geometric distribution with parameter p where $0 < p < 1$ if

- the sample space of X is the set $\{1, 2, 3, \ldots\}$
- $P(X = r) = (1 - p)^{r-1} p$

If this is the case, you can write "X is Geo(p)" as shorthand for X has the geometrical distribution with parameter p.

In the case of the random variable Y, described above, it is said that Y is Geo(0.25).

You can generalise this example to obtain guidelines for the application of the Geometric distribution:

If a simple experiment with probability p of success is repeated in a series of independent trials until the first success is obtained, and X denotes the total number of times the experiment must be conducted to achieve the first success **then**

X is Geo(p)

So, for example,

if X is the number of tosses of a fair coin required to obtain the first head then X is Geo(0.5), and if Y is the number of throws of a normal fair cubical dice to obtain the first six then Y is Geo($\frac{1}{6}$).

Experimental Verification of the Use of the Geometric Distribution

It is possible to see that Geo(1/6) is indeed a good model for the number of throws of a fair die required to obtain the first six by comparing experimental and theoretical results.

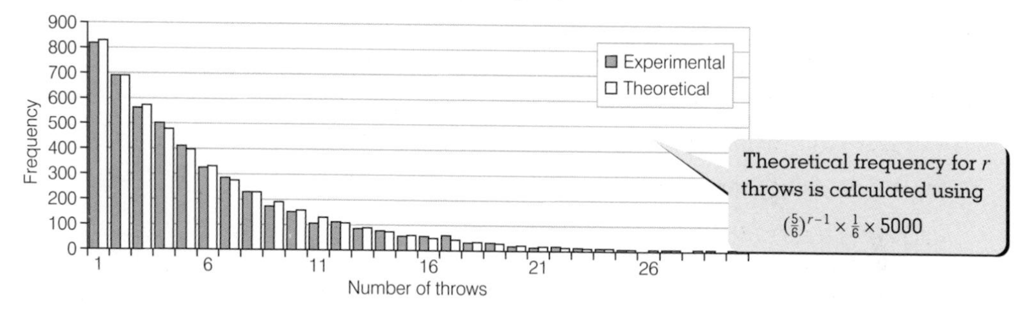

This bar graph shows a comparison of the experimental and theoretical frequencies if the experiment of counting the number of throws until a six is observed is repeated 5000 times. There is very good agreement between the two sets of frequencies and this verifies that Geo(1/6) is an appropriate model for the number of throws required for the first six.

Using the Geometric Distribution

EXAMPLE 1

Biologists have observed that 15% of a particular species of bird are albinos (i.e. completely white). A bird spotter wishes to see an albino specimen of the species. Find the probability that:

a) the third bird of the species that he sees will be the first albino;
b) he must see more than 10 birds of the species to see his first albino;
c) he will need to see less than 8 birds in order to see his first albino;
d) his first albino will be between the fifth and the fifteenth bird of the species (inclusive) that he sees.

Comment on any assumptions you have made in answering this question.

EXAMPLE 1 (continued)

SOLUTION

Let X = number of birds that must be seen to observe the first albino specimen.

X is Geo(0.15) (provided that the colour of each bird seen is independent of the colours of the other birds seen).

A_i = the ith bird seen is albino
N_i = the ith bird seen is normal

a) $P(X = 3) = P(N_1 N_2 A_3) = 0.85^2 \times 0.15 = 0.108$ (3 s.f.)

b) If he must see more than 10 birds in order to observe his first albino then **the first 10 birds must all be normal** so

$$P(X > 10) = P(N_1 N_2 N_3 \ldots N_{10}) = 0.85^{10} = 0.197 \quad \text{(3 s.f.)}$$

c) There are two different ways of answering this:

$$P(X < 8) = P(X = 1, 2, 3, \ldots, 7)$$
$$= 0.15 + 0.85 \times 0.15 + 0.85^2 \times 0.15 + \cdots + 0.85^6 \times 0.15 = 0.679 \quad \text{(3 s.f.)}$$

The calculation can either be done directly on your calculator which is quite lengthy or, if your calculator can handle sigma notation, then the probability can be written as

$$\sum_{r=1}^{7} 0.85^{r-1} \times 0.15$$

Alternatively

$$P(X < 8) = 1 - P(X \geqslant 8) = 1 - P(N_1 N_2 \ldots N_7) = 1 - 0.85^7 = 0.679 \quad \text{(3 s.f.)}$$

d) Again, this can be done in two different ways:

$$P(5 \leqslant X \leqslant 15) = P(X = 5, 6, 7, \ldots, 15)$$
$$= 0.85^4 \times 0.15 + 0.85^5 \times 0.15 + 0.85^6 \times 0.15 + \cdots + 0.85^{14} \times 0.15$$
$$= \sum_{r=5}^{15} 0.85^{r-1} \times 0.15 = 0.435 \quad \text{(3 s.f.)}$$

In practice, you would certainly want to use a graphical calculator to evaluate this!

Alternatively

$$P(5 \leqslant X \leqslant 15) = P(X \leqslant 15) - P(X \leqslant 4)$$
$$= 1 - P(X > 15) - (1 - P(X > 4))$$
$$= P(X > 4) - P(X > 15)$$
$$= 0.85^4 - 0.85^{15} = 0.435 \quad \text{(3 s.f.)}$$

Remember

$$P(X > 4) = P(\text{first 4 birds are normal}) = 0.85^4$$

and, similarly

$$P(X > 15) = P(\text{first 15 birds are normal}) = 0.85^{15}$$

When you discuss the assumptions being made in applying a particular probability model **it is important your comments take into account the context of the situation.** In this case it would not be enough just to say that independence and constant probability are being assumed.

It has been assumed that the colours of the birds seen are independent of each other: this may well not be the case if the local population being sampled from does not reflect the whole species.

Estimating the Mean of a Geometric Distribution

A biased die has probability 0.25 of landing on a six. What is the expected number of throws required to obtain the first six?

Let Y denote the number of throws required. Y is Geo(0.25)

$$\mu = \text{E}[\text{Y}] = \sum yp = \sum_{r=1}^{\infty} r \times \text{P}(\text{Y} = r) = \sum_{r=1}^{\infty} r \times 0.75^{r-1} \times 0.25$$

Unfortunately a graphical calculator will not evaluate the sum of an infinite number of terms, but you can estimate the mean of the geometric distribution by taking the sum of the first 200 terms as an **estimate** for E[Y]:

$$\mu = \text{E}[\text{Y}] = \sum_{r=1}^{\infty} r \times 0.75^{r-1} \times 0.25 \approx \sum_{r=1}^{200} r \times 0.75^{r-1} \times 0.25 = 3.99999999\ldots = 4.00 \quad (3 \text{ s.f.})$$

In a similar way, for a random variable X which is Geo(0.4), the mean can be estimated as

$$\mu = \text{E}[\text{X}] = \sum_{r=1}^{\infty} r \times 0.6^{r-1} \times 0.4 \approx \sum_{r=1}^{200} r \times 0.6^{r-1} \times 0.4 = 2.50$$

and, for a random variable W which is Geo(2/3), the mean can be estimated as

$$\mu = \text{E}[\text{W}] = \sum_{r=1}^{\infty} r \times (\tfrac{1}{3})^{r-1} \times \tfrac{2}{3} \approx \sum_{r=1}^{200} r \times (\tfrac{1}{3})^{r-1} \times \tfrac{2}{3} = 1.50$$

Observing that $4 = \frac{1}{0.25}$, $2.5 = \frac{1}{0.4}$ and $1.5 = \frac{1}{2/3}$, the following generalisation can be made

If X is Geo(p) then $\text{E}[\text{X}] = \frac{1}{p}$

This result can be proved using results from the C2 and C3 modules. See the Extension for this proof.

EXAMPLE 2

Pauline knows from experience that, when she phones Michael, on 72% of occasions Michael will be busy and unable to answer the phone. Pauline decides to make one attempt to phone Michael once each day until he answers the phone.

Let X denote the number of phone calls that Pauline must make.

a) Suggest a suitable model for X, stating two assumptions that must be made to ensure that the model is valid.

b) Calculate P(X > E[X]).

SOLUTION

a) If the success/failure of each day's attempt is independent of the earlier attempts and the probability of a successful call is always 0.28 then X is Geo(0.28).

(Note that Michael's work schedules and holiday patterns could easily make these assumptions unrealistic.)

b) If X is Geo(0.28) then $\text{E}[\text{X}] = \frac{1}{0.28} = \frac{100}{28} = \frac{25}{7}$

$$\begin{aligned} \text{P}(\text{X} > \text{E}[\text{X}]) &= \text{P}(\text{X} > \tfrac{25}{7}) = \text{P}(\text{X} = 4, 5, 6, \ldots) \\ &= \text{P(Michael is unavailable for first three calls)} \\ &= 0.72^3 = 0.373 \quad (3 \text{ s.f.}) \end{aligned}$$

EXERCISE 1

1 A fair coin is tossed repeatedly until a head is obtained. Find the probability that:

a) exactly three tosses are required;
b) more than three tosses are required;
c) less than five tosses are required;
d) between two and six (inclusive) tosses are required.

2 A fair die is thrown repeatedly until a 3 is obtained. Find the probability that:

a) exactly four throws are required;
b) more than ten throws are required;
c) less than eight throws are required;
d) between two and nine (inclusive) throws are required.

3 A card is drawn from a pack, its suit noted, the card replaced and the pack thoroughly reshuffled. This is repeated until a HEART is obtained. Find the probability that:

a) exactly four draws are required;
b) at least seven draws are required;
c) at most ten draws are required;
d) between two and nine (exclusive) draws are required.

4 An angler knows that each time she casts her line into a river she has a 1 in 20 chance of catching a fish. She is interested in how many times she will need to cast her line in order to catch a fish.

a) Stating any necessary assumption, name an appropriate distribution with which to model this situation. What is the expected number of times she will need to cast her line in order to catch a fish.
b) Find the probability that she will catch her first fish the fourth time that she casts her line.
c) Find the probability that she will need to cast her line more than 12 times in order to catch her first fish.

5 An archer has probability 0.15 of hitting the bulls-eye with each arrow that she fires. During a training session she fires a large number of arrows at the target. Let X denote the number of bulls-eyes she obtains in her first 20 attempts and let Y denote the number of arrows she must fire to obtain the first bulls-eye.

a) i) Describe a suitable model for the random variable X. State, and comment on, any assumptions that must be made for this model to be appropriate.
ii) Calculate $P(X \leqslant E[X])$.
b) i) Describe a suitable model for the random variable Y. State, and comment on, any assumptions that must be made for this model to be appropriate.
ii) Calculate $P(Y > E[Y])$.

6 Tea manufacturers often used to promote brand loyalty by putting cards into their packs of tea. There was usually a series of 50 cards to collect on a theme such as "Wild Animals of the World" or "Veteran Cars". Without opening a packet, it is impossible to tell which card the packet contains.
Thomas is collecting a series of these cards. He currently has 40 cards in the series. Let X_{41} denote the number of packets of tea that Thomas must buy in order to get one of the ten remaining cards he needs for his collection.

a) Stating any necessary assumption, name an appropriate distribution with which to model X_{41}. What is the expected number of packets he must buy in order to get one of the remaining cards?

b) What is the probability that he needs to buy four packets in order to get a new card?
c) What is the probability that he will need to buy at most six packets in order to get a new card?
d) What is the probability that he will need to buy at least ten packets in order to get a new card?

7 Sarah is a javelin thrower. She knows that with each throw she has probability 0.17 of throwing the javelin further than 55 m. During a training session, she repeatedly throws the javelin until a throw exceeds 55 m. Let U be the number of throws required.
a) State a possible probability model for U. State two assumptions, in context, that must be made for this model to be appropriate. How reasonable are these assumptions?
b) Calculate:
i) $P(U = 4)$;
ii) $P(U > 6)$;
iii) $P(U < 4)$;
iv) $P(5 \leqslant U \leqslant 10)$.

In an athletics competition, each competitor throws the javelin six times. Let V denote the number of Sarah's throws that exceed 55 m.
c) State a possible probability model for V.
d) Calculate $P(V = 2)$.

8 The independent random variables X and Y are Geo(0.625) and B(4, p), respectively. The two random variables have equal means.
a) Calculate the value of p.
b) Calculate the probability that X and Y both equal 2.
c) Calculate $P(X = Y)$.

EXTENSION

Outline of the Proof of the E[X] Result for the Geometric Distribution

Suppose X is Geo(p) then $E[X] = \sum_{r=1}^{\infty} rP(X = r) = \sum_{r=1}^{\infty} r \times (1-p)^{r-1} p$

Now consider

$$S(x) = 1 + x + x^2 + x^3 + x^4 + \cdots$$

This is the sum to infinity of a geometric progression with common ratio x.

If $-1 < x < 1$ then you can write

$$S(x) = 1 + x + x^2 + x^3 + x^4 + \cdots = \frac{1}{1-x} = (1-x)^{-1}$$

Differentiating this equation (using the chain rule for the right-hand side) gives

$$S'(x) = 1 + 2x + 3x^2 + 4x^3 + \cdots = (1-x)^{-2}$$

If you put $x = 1 - p$ you obtain

$$1 + 2(1-p) + 3(1-p)^2 + 4(1-p)^3 + \cdots = (1-(1-p))^{-2}$$

$$\Rightarrow \quad 1 + 2(1-p) + 3(1-p)^2 + 4(1-p)^3 + \cdots = p^{-2}$$

and multiplying through by p gives

$$p + 2p(1-p) + 3p(1-p)^2 + 4p(1-p)^3 + \cdots = p \times p^{-2}$$

$$\Rightarrow \quad \sum_{r=1}^{\infty} rp(1-p)^{r-1} = p^{-1}$$

Therefore

$$E[X] = \sum_{r=1}^{\infty} rP(X = r) = \sum_{r=1}^{\infty} r \times (1-p)^{r-1} p = p^{-1}$$

so

$$\text{If X is Geo}(p) \text{ then } E[X] = \frac{1}{p}$$

Having studied this chapter you should know that

- a random variable X is said to have the geometric distribution with parameter p where $0 < p < 1$
  if X takes the values 1, 2, 3, ...
  and $P(X = r) = (1 - p)^{r-1}p$
  if this is the case, you write "X is Geo(p)" as shorthand for X has the geometric distribution with parameter p
- **if** a simple experiment with probability p of success is repeated in a series of independent trials until the first success is obtained, and X denotes the total number of times the experiment must be conducted to achieve the first success **then**
  X is Geo(p)
- if X is Geo(p) then $E[X] = \frac{1}{p}$

REVISION EXERCISE

1 A footballer has probability 0.19 of scoring a goal when he has a shot. Let T denote the number of shots at goal that the footballer makes at the beginning of a new football season up to and including the shot when he scores his first goal.

a) State two assumptions, in context, that must be made if the Geo(0.19) distribution is to be a good probability model for T.

b) Use the Geo(0.19) distribution to find:

i) $P(T = 3)$;

ii) $P(T > 8)$;

iii) $P(T < E[T])$.

2 The random variable X is Geo(0.7) and the random variable Y is B(4, 0.5). Find:

a) $P(X = 2 \text{ or } 3)$;

b) $P(Y = 1)$.

If X and Y are independent, find:

c) $P(X = Y)$.

3 Every time a darts player throws a dart at a dartboard, the probability that she scores a "double" is $\frac{1}{8}$ independently of the result of any other throw. She keeps throwing until she scores a double. Let T be the number of throws taken by the player up to and including the throw with which she first scores a double.

a) Name the distribution of T and find E[T].

b) Find $P(T = 3)$

(OCR Jun 2001 S1)

4 Adill and Beth are playing a game. Adill throws a fair die three times. The number of sixes she obtains is denoted by A. Beth throws a fair coin repeatedly. The number of throws up to and including the first throw on which the coin lands head upwards is denoted by B.

i) State the distribution of A, giving the values of any parameters.

ii) State the distribution of B, giving the values of any parameters.

iii) Find $P(A = 2)$.

iv) Find $P(B > 2)$.

v) Find $P(A = B)$.

(OCR Jan 2002 S1)

5 The probability that a ticket in a national lottery is a prize winning ticket is 0.02. Peter buys one lottery ticket each week for two years (104 weeks). Let W be the number of winning tickets that he obtains.

a) State the distribution of W.

b) Find $P(W = 2)$.

c) Find the mean and standard deviation of W.

d) Find $P(W < E[W])$.

Hayley buys one ticket each week until she buys a winning ticket. Let H be the number of tickets that Hayley buys.

e) State the distribution of H.

f) Find $P(H > 52)$.

g) Find E[H].

6 On a fairground stall the prize is a large cuddly toy worth £20. In order to win this prize Vicky needs to throw a hoop over a pole. For each throw, her probability of success is $\frac{2}{35}$, independently of all other throws. She pays the owner of the stall 50p for each throw and keeps throwing the hoop until she wins the prize.

a) Calculate the probability that Vicky takes fewer than 3 throws to win the prize.

b) Find the expected number of throws that Vicky needs to win the prize.

c) Calculate the probability that Vicky pays more to win the prize than it is worth.

(OCR May 2002 S1)

7 The random variable G has a geometric distribution with expectation 9. Find $P(G = 4)$ and $P(G \leqslant 7)$.

(OCR Jun 2000 S1, part)

8 It is given that one-eleventh of adults are left-handed. Adults are chosen, one by one, until a left-handed adult is found. The total number of adults chosen, including the left-handed one, is denoted by T. State an assumption that must be made for a geometric distribution to be a suitable model for T.
Using a geometric distribution:
i) calculate $P(T = 4)$;
ii) calculate $P(T > 10)$;
iii) find the mean of T.

(OCR Jun 1999 S1, amended)

9 The random variable V has a geometric distribution and takes the values 1, 2, 3, ...
It is given that $P(V = 1) = 0.32$.
a) Calculate $P(V = 4)$.
b) Find the expected value of V.

(OCR Mar 2000 S1, part)

10 An experiment is repeated until it is successful. The experiments are independent of each other and each has the same probability p of being successful. The number of experiments required, including the successful one, is denoted by L.
a) Given that $p = 0.7$, find:
  i) $P(L = 4)$;
  ii) the mean value of L.
b) **i)** Express, in terms of p, the probability that the first four experiments are all unsuccessful.
  ii) Hence, given that $P(L > 4) = 0.1296$, find the value of p.

(OCR Nov 1999 S1, amended)

11 The manufacturers of brand X catfood plan a television advertisement based on the claim that "Eight cats out of ten prefer brand X". In fact brand X is indistinguishable from any other brand, so that, given a choice between brand X and another brand, the probability that any cat chooses brand X is 0.5. Ten cats are selected at random and each is given a choice between brand X and another brand. Show that the probability that 8 or more choose brand X is 0.0547, correct to 3 significant figures.

In order to achieve a satisfactory result, the manufacturers of brand X select groups of 10 cats at random until 8 or more out of a group of 10 cats choose brand X. The number of groups selected, including the successful group, is denoted by Y. Calculate:
i) the probability that $Y = 4$;
ii) the probability that $Y > 6$;
iii) the mean of Y.

(OCR Jun 1997 S1, amended)

10 Correlation and Regression

The purpose of this chapter is to enable you to

- use scatter diagrams to illustrate bivariate data
- use the product moment correlation coefficient as a measure of how closely data points may be approximated by a straight line
- make valid conclusions from a calculated correlation coefficient
- calculate the equations of least squares regression lines

Bivariate Data

So far in this module, just one piece of information has been taken from each member of the sample: for example the travelling times of 1000 commuters or the weight of 100 three-month old babies.

Very often, members of a sample are asked to give more than one piece of information: for example, the travelling times and distances of 1000 commuters or the height, weight and IQ of 100 thirteen-year old girls. Such data is called **multivariate data**.

Experiments in which exactly two pieces of data are gathered from each member of the sample are called **bivariate experiments**.

Consider the following three sets of bivariate data.

Data set 1: This shows the highest and lowest values (in pence) of the shares of 10 electronics companies for the 12-month period May 2003–April 2004.

High	355	63	202	368	40	329	171	247	95	287
Low	200	31	143	211	14	167	102	132	37	169

Data set 2: This shows the prices and mileages of 13 used 1800 cc cars advertised for sale during April 2004.

Mileage	54 000	41 000	50 000	44 000	39 000	35 000	31 000	20 000	25 000	34 000	18 000	24 000	12 000
Price	4499	4699	4908	4994	4995	5299	5495	5495	5695	5795	5999	6399	6495

Data set 3: This shows the percentage marks in Maths and Art of 10 students randomly selected from a year 9 class.

Pupil	1	2	3	4	5	6	7	8	9	10
Maths %	80	48	62	53	45	72	60	41	53	57
Art %	53	61	56	72	48	51	45	58	50	46

Whenever you have bivariate data the question of whether there is any relationship between the two sets of data naturally arises. It is possible to establish diagrammatic procedures to determine whether there is a relationship and arithmetic procedures to decide whether a **linear** relationship exists.

Scatter Diagrams

For bivariate data a **scatter diagram** is usually drawn to display the data in a visual form. Each member of the sample gives a point on the graph where the first piece of data determines the x co-ordinate and the second piece of data determines the y co-ordinate.

It is a good idea also to mark in the point whose x value is the mean of the first pieces of data and whose y value is the mean of the second pieces of data. This is the point $(\bar{x}, \bar{y})$.

Here is the scatter diagram for the share prices of data set 1:

											Means
High	355	63	202	368	40	329	171	247	95	287	*215.7*
Low	200	31	143	211	14	167	102	132	37	169	*120.6*

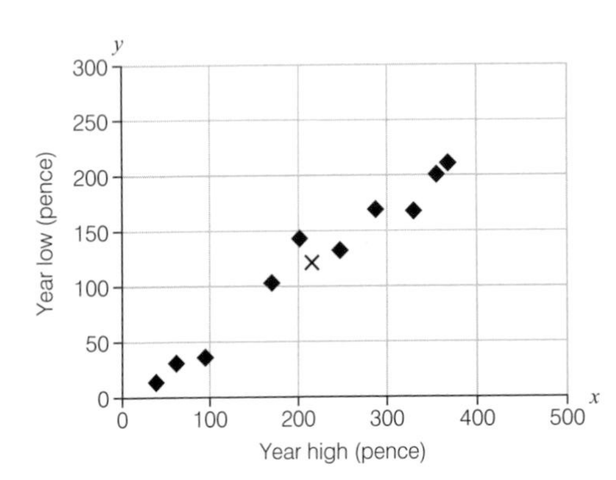

The diamonds correspond to the data points whilst the cross represents the point $(\bar{x}, \bar{y})$.

In this case there appears to be a link between the year highs and year lows of the shares.

The data are quite tightly packed about a line of positive gradient: we say that the data exhibit **high positive linear correlation**.

Here is the scatter diagram for the car mileages and prices of data set 2:

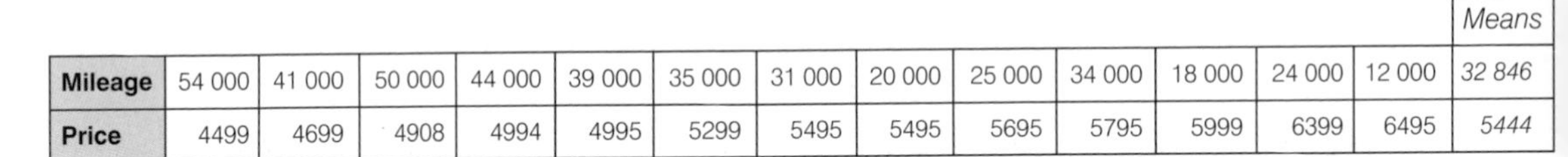

													Means
Mileage	54 000	41 000	50 000	44 000	39 000	35 000	31 000	20 000	25 000	34 000	18 000	24 000	12 000	*32 846*
Price	4499	4699	4908	4994	4995	5299	5495	5495	5695	5795	5999	6399	6495	*5444*

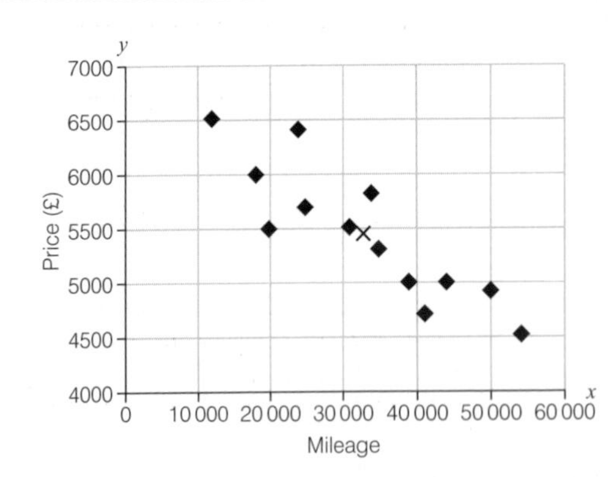

Again there seems to be a link: the data are reasonably tightly packed about a line of negative gradient.

It can be said that the data exhibit **negative linear correlation**.

In cases where there seems to be a linear pattern but the data are not as tightly packed about a line it can be said that there is weak linear correlation.

Here is the scatter diagram for the exam marks of data set 3:

Pupil	1	2	3	4	5	6	7	8	9	10	*Mean*
Maths %	80	48	62	53	45	72	60	41	53	57	*57.1*
Art %	53	61	56	72	48	51	45	58	50	46	*54*

There is no obvious relationship between the Art marks and the Maths marks of the class: it can be said that the variables are **uncorrelated**.

Now consider this scatter diagram:

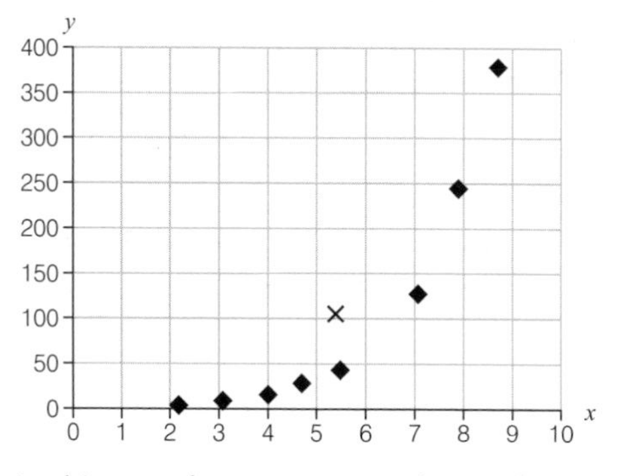

In this case there appears to be a relationship between the two variables but it is certainly not a linear relationship.

The Product Moment Coefficient of Linear Correlation

A measure of how closely a set of data points clusters around a line is given by r, **the product moment coefficient of linear correlation**. This is sometimes called Pearson's correlation coefficient or just the correlation coefficient of the data.

For data points $(x_1, y_1), (x_2, y_2), \ldots, (x_n, y_n)$ the product moment coefficient of linear correlation, r, is given by

$$r = \frac{S_{xy}}{\sqrt{S_{xx}S_{yy}}}$$

where

$$S_{xx} = \Sigma (x_i - \bar{x})^2 = \Sigma x_i^2 - \frac{(\Sigma x_i)^2}{n}, \quad S_{yy} = \Sigma (y_i - \bar{y})^2 = \Sigma y_i^2 - \frac{(\Sigma y_i)^2}{n}$$

$$S_{xy} = \Sigma (x_i - \bar{x})(y_i - \bar{y}) = \Sigma x_i y_i - \frac{(\Sigma x_i)(\Sigma y_i)}{n}$$

An argument motivating the formula for the correlation coefficient is presented in Extension 1 to this chapter.

All these formulae are given in the OCR formula book.

Calculating the Correlation Coefficient

The product moment correlation coefficient for the share price data of data list 1 can easily be calculated:

High price x **pence**	355	63	202	368	40	329	171	247	95	287
Low price y **pence**	200	31	143	211	14	167	102	132	37	169

The routine details of the calculation can be presented in a table or done on a calculator:

High (x)	Low (y)	x^2	y^2	xy
355	200	126 025	40 000	71 000
63	31	3969	961	1953
202	143	40 804	20 449	28 886
368	211	135 424	44 521	77 648
40	14	1600	196	560
329	167	108 241	27 889	54 943
171	102	29 241	10 404	17 442
247	132	61 009	17 424	32 604
95	37	9025	1369	3515
287	169	82 369	28 561	48 503
2157	1206	597 707	191 774	337 054

and the correlation coefficient can then be calculated:

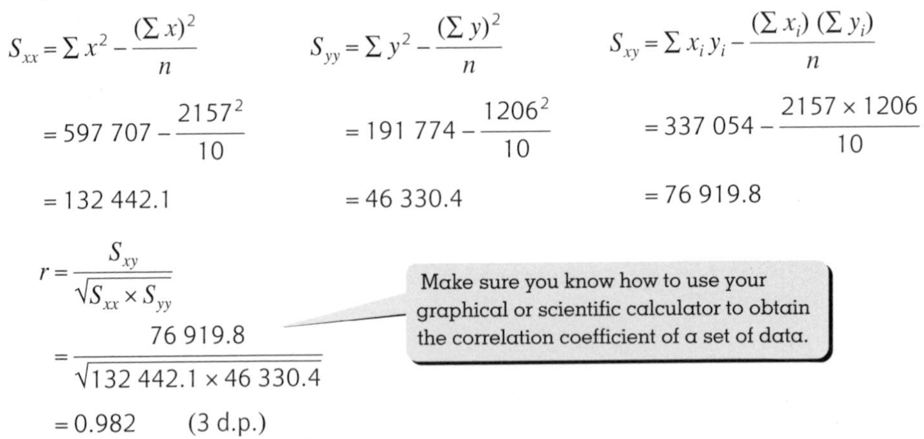

$n = 10$

$$S_{xx} = \Sigma x^2 - \frac{(\Sigma x)^2}{n} = 597\,707 - \frac{2157^2}{10} = 132\,442.1$$

$$S_{yy} = \Sigma y^2 - \frac{(\Sigma y)^2}{n} = 191\,774 - \frac{1206^2}{10} = 46\,330.4$$

$$S_{xy} = \Sigma x_i y_i - \frac{(\Sigma x_i)(\Sigma y_i)}{n} = 337\,054 - \frac{2157 \times 1206}{10} = 76\,919.8$$

$$r = \frac{S_{xy}}{\sqrt{S_{xx} \times S_{yy}}} = \frac{76\,919.8}{\sqrt{132\,442.1 \times 46\,330.4}} = 0.982 \quad \text{(3 d.p.)}$$

Calculating the Correlation Coefficient from Summarised Data

Sometimes the data may be summarised for you so that the calculation of the correlation coefficient is merely a question of applying the formulae correctly.

For example, if for the exam mark data of data set 3 you were told that

$$n = 10, \sum x = 571, \sum y = 540, \sum x^2 = 33\,905, \sum y^2 = 29\,760, \sum xy = 30\,638$$

then you could use the initial formula for correlation to calculate the value of r:

The values above give

$$S_{xx} = \sum x^2 - \frac{(\sum x)^2}{n} = 33\,905 - \frac{571^2}{10} = 1300.9$$

$$S_{yy} = \sum y^2 - \frac{(\sum y)^2}{n} = 29\,760 - \frac{540^2}{10} = 600$$

$$S_{xy} = \sum xy - \frac{(\sum x)(\sum y)}{n} = 30\,638 - \frac{571 \times 540}{10} = -196$$

and

$$r = \frac{S_{xy}}{\sqrt{S_{xx}S_{yy}}} = \frac{-196}{\sqrt{1300.9 \times 600}} = -0.222$$

Properties of the Correlation Coefficient

The table below shows the data values, scatter graphs and the values of the correlation coefficient for five sets of data.

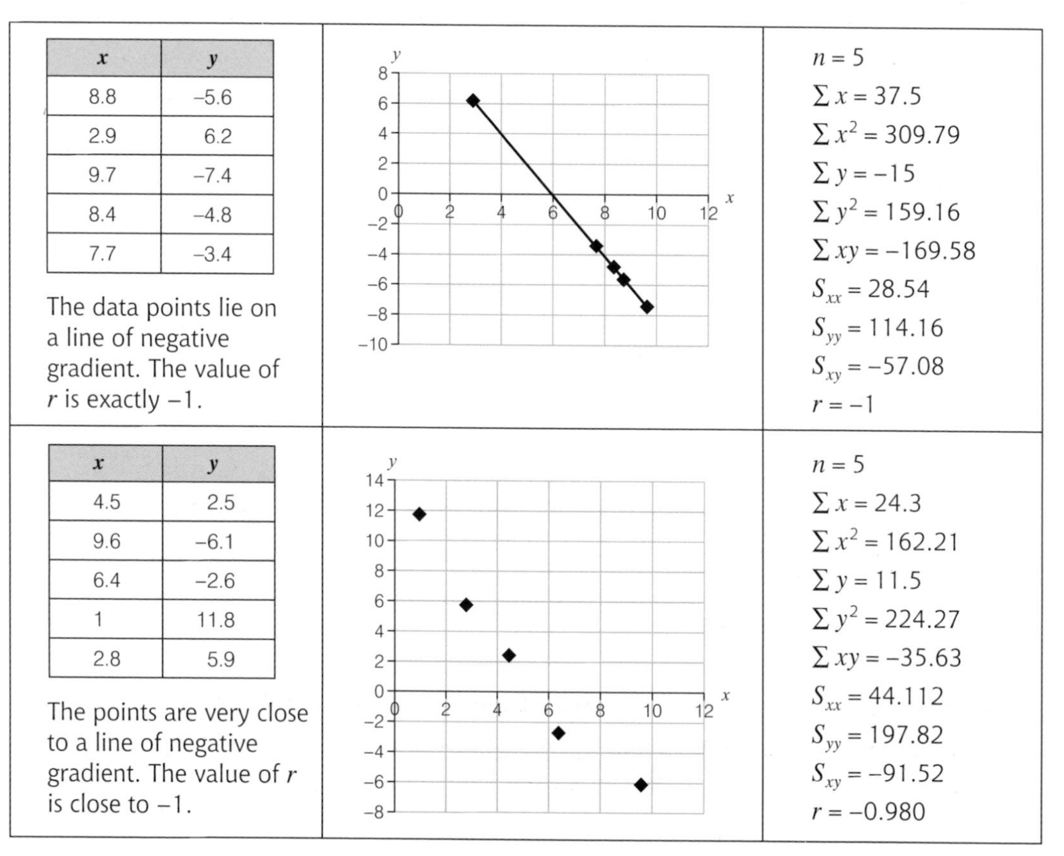

x	y
8.8	−5.6
2.9	6.2
9.7	−7.4
8.4	−4.8
7.7	−3.4

The data points lie on a line of negative gradient. The value of r is exactly −1.

$n = 5$
$\sum x = 37.5$
$\sum x^2 = 309.79$
$\sum y = -15$
$\sum y^2 = 159.16$
$\sum xy = -169.58$
$S_{xx} = 28.54$
$S_{yy} = 114.16$
$S_{xy} = -57.08$
$r = -1$

x	y
4.5	2.5
9.6	−6.1
6.4	−2.6
1	11.8
2.8	5.9

The points are very close to a line of negative gradient. The value of r is close to −1.

$n = 5$
$\sum x = 24.3$
$\sum x^2 = 162.21$
$\sum y = 11.5$
$\sum y^2 = 224.27$
$\sum xy = -35.63$
$S_{xx} = 44.112$
$S_{yy} = 197.82$
$S_{xy} = -91.52$
$r = -0.980$

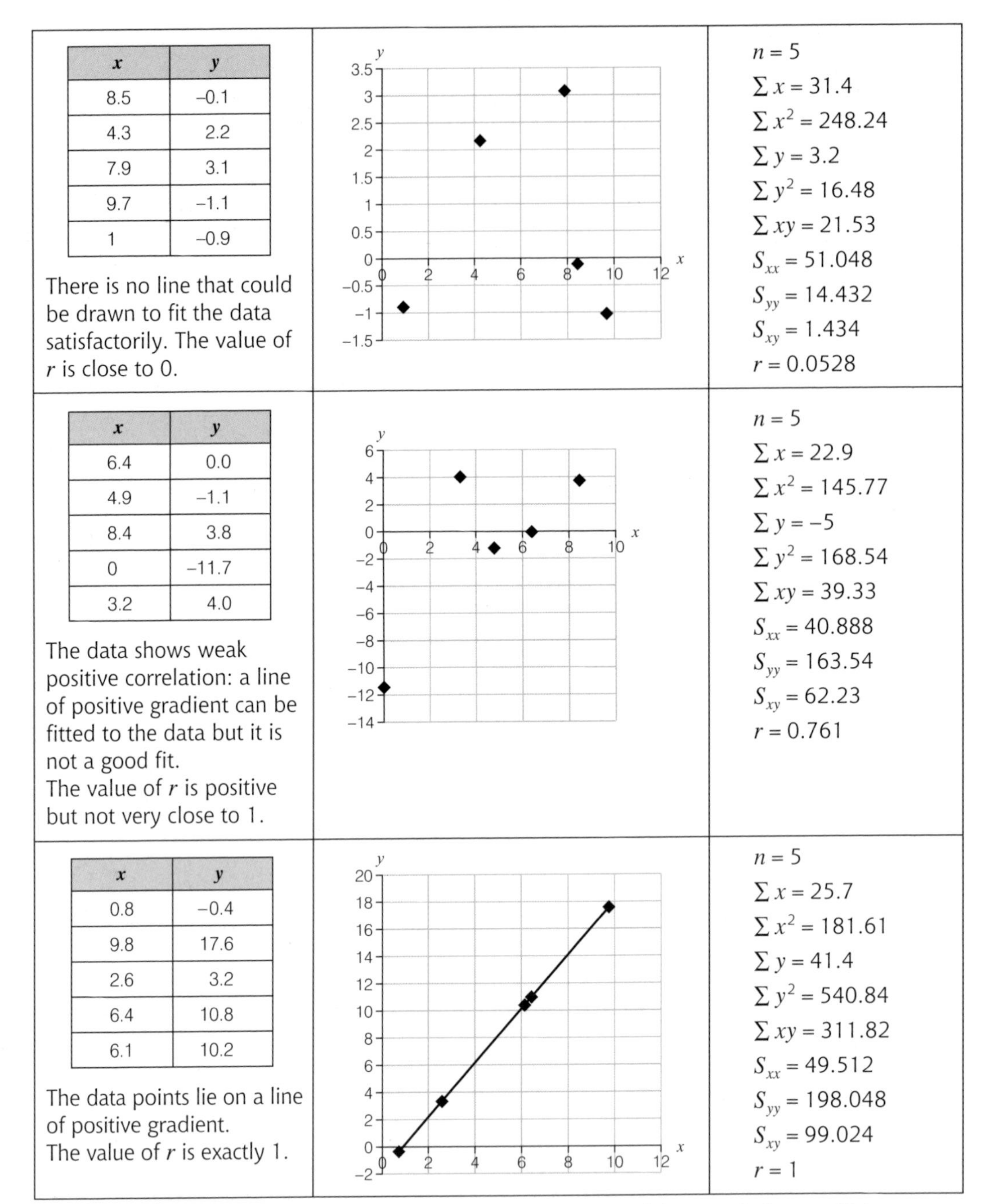

x	y
8.5	–0.1
4.3	2.2
7.9	3.1
9.7	–1.1
1	–0.9

There is no line that could be drawn to fit the data satisfactorily. The value of r is close to 0.

$n = 5$
$\sum x = 31.4$
$\sum x^2 = 248.24$
$\sum y = 3.2$
$\sum y^2 = 16.48$
$\sum xy = 21.53$
$S_{xx} = 51.048$
$S_{yy} = 14.432$
$S_{xy} = 1.434$
$r = 0.0528$

x	y
6.4	0.0
4.9	–1.1
8.4	3.8
0	–11.7
3.2	4.0

The data shows weak positive correlation: a line of positive gradient can be fitted to the data but it is not a good fit.
The value of r is positive but not very close to 1.

$n = 5$
$\sum x = 22.9$
$\sum x^2 = 145.77$
$\sum y = -5$
$\sum y^2 = 168.54$
$\sum xy = 39.33$
$S_{xx} = 40.888$
$S_{yy} = 163.54$
$S_{xy} = 62.23$
$r = 0.761$

x	y
0.8	–0.4
9.8	17.6
2.6	3.2
6.4	10.8
6.1	10.2

The data points lie on a line of positive gradient.
The value of r is exactly 1.

$n = 5$
$\sum x = 25.7$
$\sum x^2 = 181.61$
$\sum y = 41.4$
$\sum y^2 = 540.84$
$\sum xy = 311.82$
$S_{xx} = 49.512$
$S_{yy} = 198.048$
$S_{xy} = 99.024$
$r = 1$

These examples illustrate most of the important properties of the correlation coefficient:

- $-1 \leqslant r \leqslant 1$
  $r = 1$ if and only if all the points lie on a straight line of positive gradient
  $r = -1$ if and only if all the points lie on a straight line of negative gradient
- if r is close to +1 or –1 then there is a good straight line fit to the data. If r is close to 0 then there is not a good linear fit to the data
- r has no units

A justification of these results can be found in Extension 1.

EXERCISE 1

1 Calculate the product moment correlation coefficient of the prices and mileages of 13 used 1800 cc cars advertised for sale during April 2004.
(Data set 2 from page 149)

Mileage	54 000	41 000	50 000	44 000	39 000	35 000	31 000	20 000	25 000	34 000	18 000	24 000	12 000
Price	4499	4699	4908	4994	4995	5299	5495	5495	5695	5795	5999	6399	6495

Comment on your answer with reference to the scatter diagram for the data presented on page 150.

For the sets of data in questions 2–6, draw a scatter diagram to illustrate the data; calculate the product moment correlation coefficient for the data and comment on the relationship between the x and y values.

2 The price and lifetime of 10 different brands of electric lightbulbs were measured:

Price (pence) x	12	18	27	22	42	61	53	39	53	35
Lifetime (hours) y	178	234	255	282	377	577	521	442	487	365

3 A survey was conducted at an office into the value of a person's housing and their commuting time:

Value, x (thousands of £)	128	166	182	90	334	196	332	480	358	420
Daily commuting time, y (minutes)	160	140	150	210	90	110	100	50	40	70

4 The table shows the gestation period (x days) and the maximum recorded life span (y years) of different species of animal:

Animal	chimp	dog	hedgehog	hippo	horse	lion	panda	pig	rabbit	squirrel
Gestation period x days	235	60	30	240	350	110	138	115	30	40
Maximum life span y years	56	29	16	54	62	29	27	27	18	23

5 The goals scored by (x) and against (y) the Premier League football clubs during the 2003–4 football season:

	Arsenal	Chelsea	Man U.	L'pool	N'cstle	Villa	Charlton	Bolton	Fulham	B'ham
x	73	67	64	55	52	48	51	48	52	43
y	26	30	35	37	40	44	51	56	46	48
	M'boro	Soton	P'mouth	Spurs	B'burn	Man C	Everton	Leicester	Leeds	Wolves
x	44	44	47	47	51	55	45	48	40	38
y	52	45	54	57	59	54	57	65	79	77

6 Two ballet judges were asked to rank ten dancers from 1–10 with 1 being the best:

Dancer	A	B	C	D	E	F	G	H	I	J
Judge 1	7	9	3	6	2	1	10	4	8	5
Judge 2	9	10	1	8	3	2	7	5	6	4

7 A sample consists of 20 data points (x_i, y_i) and can be summarised by

$$\sum x = 210, \sum x^2 = 2870, \sum y = 1163, \sum y^2 = 73\,396.66, \sum xy = 14\,140$$

Calculate the value of r.

8 A sample of 50 data points (x_i, y_i) can be summarised by

$$\sum (x - \bar{x})^2 = 37.82, \sum (y - \bar{y})^2 = 85.93, \sum (x - \bar{x})(y - \bar{y}) = -45.56$$

Calculate the value of r.

For each of the sets of data in questions 9–13:

a) use a calculator or the given summary values to calculate the product moment correlation coefficient for the data;

b) use your answer together with the scatter diagram, which has already been drawn, to comment briefly on the relationship between the x and y values.

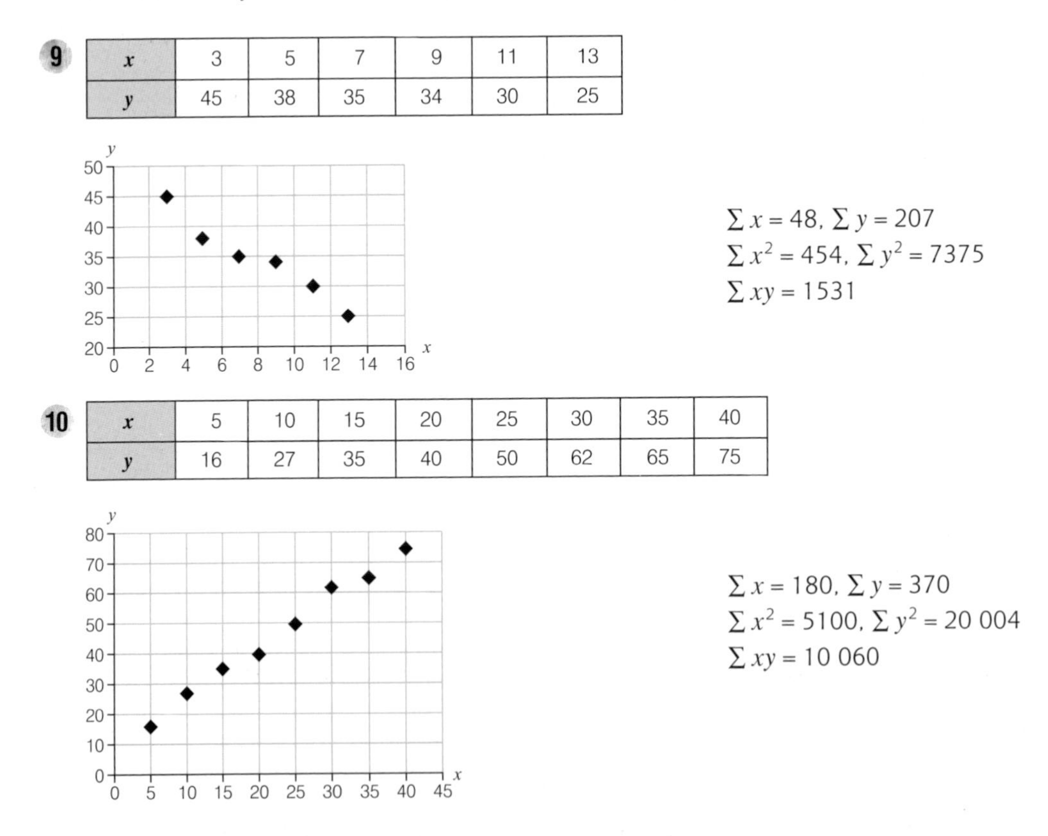

9

x	3	5	7	9	11	13
y	45	38	35	34	30	25

$\sum x = 48, \sum y = 207$

$\sum x^2 = 454, \sum y^2 = 7375$

$\sum xy = 1531$

10

x	5	10	15	20	25	30	35	40
y	16	27	35	40	50	62	65	75

$\sum x = 180, \sum y = 370$

$\sum x^2 = 5100, \sum y^2 = 20\,004$

$\sum xy = 10\,060$

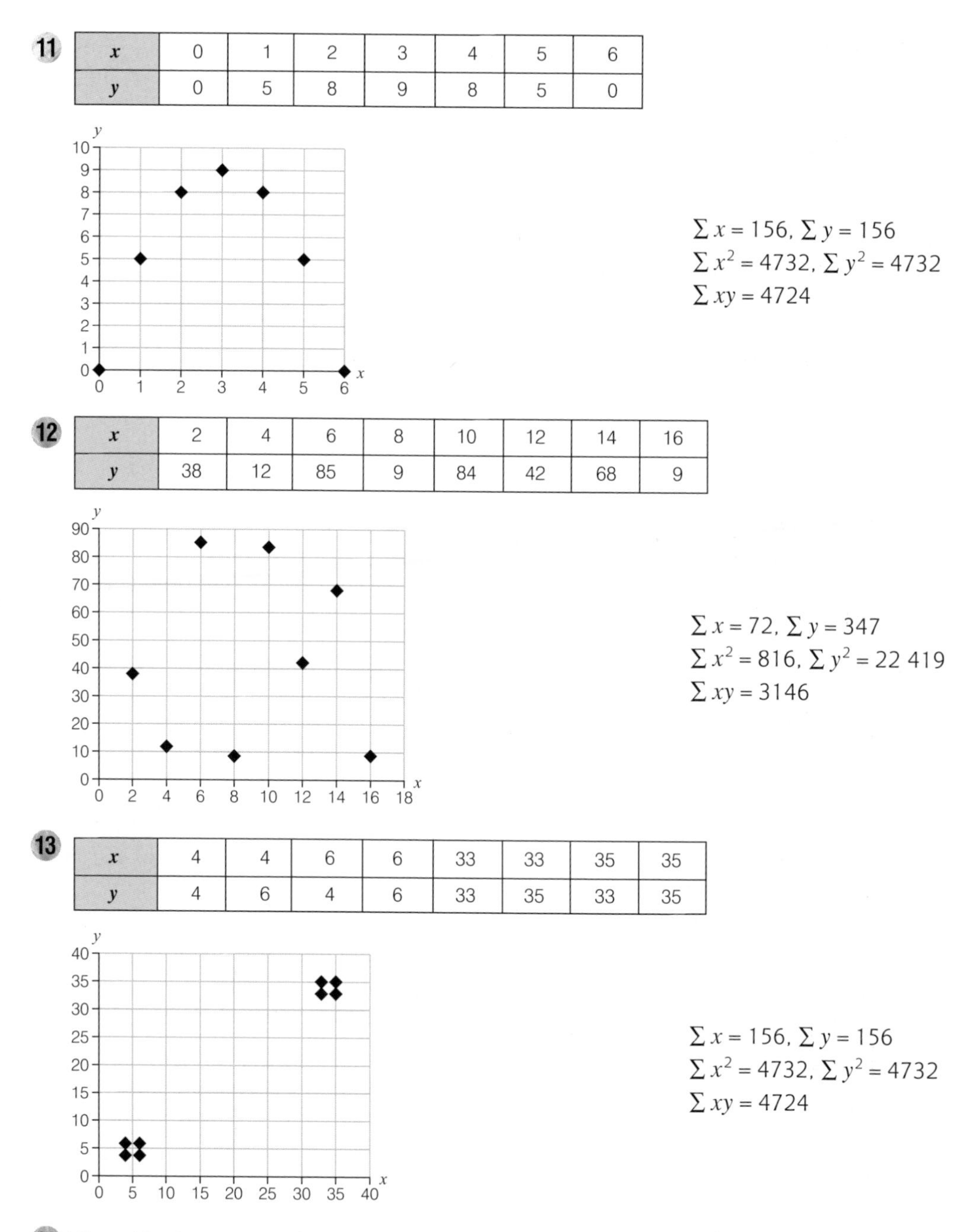

11

x	0	1	2	3	4	5	6
y	0	5	8	9	8	5	0

$\sum x = 156, \sum y = 156$
$\sum x^2 = 4732, \sum y^2 = 4732$
$\sum xy = 4724$

12

x	2	4	6	8	10	12	14	16
y	38	12	85	9	84	42	68	9

$\sum x = 72, \sum y = 347$
$\sum x^2 = 816, \sum y^2 = 22\,419$
$\sum xy = 3146$

13

x	4	4	6	6	33	33	35	35
y	4	6	4	6	33	35	33	35

$\sum x = 156, \sum y = 156$
$\sum x^2 = 4732, \sum y^2 = 4732$
$\sum xy = 4724$

14 The table shows some data values (x_i, y_i) collected in an experiment:

x	3.2	4.3	5.7	6.8	7.4	8.1
y	10.8	14.1	19.2	21.5	24.5	25.2

a) Calculate the product moment correlation coefficient for this data.

New data values (u_i, v_i) are obtained from these data values using the rules

$u_i = 2x_i - 1$ and $v_i = 0.3y_i + 2$

The new data values are shown in the table below:

u	5.4	7.6	10.4	12.6	13.8	15.2
v	5.24	6.23	7.76	8.45	9.35	9.56

b) Calculate the product moment correlation coefficient for the new data values.
c) Comment on your answers to (a) and (b).

Drawing Conclusions from Calculated Correlation Coefficients

It is important to take care when drawing conclusions from values of the product moment correlation coefficient. It must always be remembered that all it does is give a measure of how close the data points are to lying on a line.

The conclusions that can be drawn are as follows.

$r = 1$	**All the data points lie on a line of positive gradient.**
r close to 1	The data points are closely packed around a line of positive gradient.
r close to 0	There is no **straight line** approximating the data points.
r close to -1	The data points are closely packed around a line of negative gradient.
$r = -1$	**All the data points lie on a line of negative gradient.**

When a correlation coefficient has been calculated for two variables care must be taken to avoid misleading conclusions.

If the correlation coefficient is close to 1 or -1 then it simply suggests that there may be a linear relationship between the two variables.

Care must be taken to avoid suggesting that a value of r close to 1 or -1 implies causation.

The fact that the weight and height of babies have a high correlation coefficient does **not** mean that a heavy baby is heavy because it is tall.

Similarly, there would certainly be a high correlation coefficient between the average number of GCSE grades A–C obtained by 16-year olds over each of the last 10 years and the average wine consumption of adults over the last 10 years, but it would be ridiculous to suggest that the wine consumption of adults was affecting the GCSE performance of 16-year olds.

There would also be a high positive correlation coefficient for data showing the average expenditure of households on holidays each year for the last 10 years and the percentage of households owning a DVD recorder. It would, though, be ridiculous to suggest that more households own a DVD now because people are spending more on holidays. In this case the positive correlation is due to a third factor: the general increase in the standard of living.

If the correlation coefficient is close to 0 then you should simply conclude that the data is not closely packed around a straight line of positive or negative gradient.

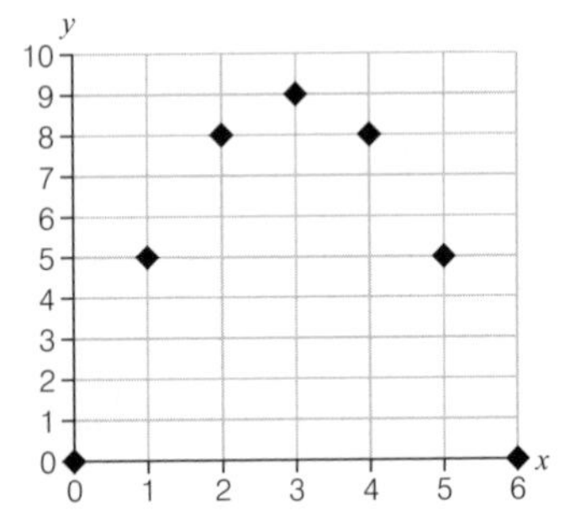

For example, the data shown in the diagram have a correlation coefficient of 0. Although the points are not close to a straight line, they all lie on the curve $y = 6x - x^2$.

Thus, variables linked by a clear rule may have correlation coefficients very close to, or even precisely, zero.

This should highlight the wisdom of drawing at least a sketch of the scatter diagram whenever a correlation coefficient is being calculated.

Combining two unlike populations can often create very high correlation coefficients which are in fact meaningless.

For example, the data shown in the diagram have a correlation coefficient of 0.995. Inspection of the scatter diagram suggests that the data possibly consists of two separate populations which have been erroneously combined.

This again highlights the importance of drawing a scatter diagram.

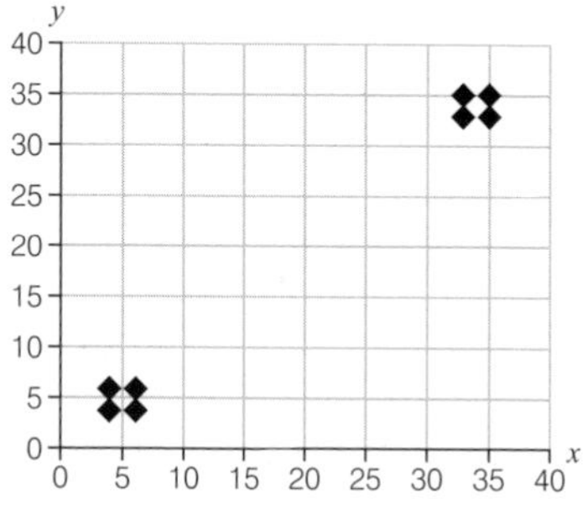

Effect of Coding on the Correlation Coefficient

Consider the data points (x, y) given by the table

x	3.2	4.7	6.1	8.2	12.4
y	21.6	17.5	14.6	10.9	4.2

and suppose the original data points are coded using

$$u = 2x - 5\ ,\ v = 5x - 20$$

to obtain new a new table of data points (u, v)

u	1.4	4.4	7.2	11.4	19.8
v	88	67.5	53	34.5	1

For the (x, y) data you have

$$n = 5,\ \Sigma x = 34.6,\ \Sigma x^2 = 290.54,\ \Sigma y = 68.8,\ \Sigma y^2 = 1122.42,\ \Sigma xy = 381.89$$

so

$$S_{xx} = \Sigma x^2 - \frac{(\Sigma x)^2}{n} = 290.54 - \frac{34.6^2}{5} = 51.108$$

$$S_{yy} = \Sigma y^2 - \frac{(\Sigma y)^2}{n} = 1122.42 - \frac{68.8^2}{5} = 175.732$$

$$S_{xy} = \Sigma xy - \frac{(\Sigma x)(\Sigma y)}{n} = 381.89 - \frac{34.6 \times 68.8}{5} = -94.206$$

and

$$r = \frac{S_{xy}}{\sqrt{S_{xx}S_{yy}}} = \frac{-94.206}{\sqrt{51.108 \times 175.732}} = -0.994 \quad \text{(to 3 d.p.)}$$

For the (u, v) data you have

$$n = 5,\ \sum u = 44.2,\ \sum u^2 = 595.16,\ \sum v = 244,\ \sum v^2 = 16300.5,\ \sum uv = 1214.9 \quad \text{so}$$

$$S_{uu} = \sum u^2 - \frac{(\sum u)^2}{n} = 595.16 - \frac{44.2^2}{5} = 204.432$$

$$S_{vv} = \sum v^2 - \frac{(\sum v)^2}{n} = 16300.5 - \frac{244^2}{5} = 4393.3$$

$$S_{xy} = \sum uv - \frac{(\sum u)(\sum v)}{n} = 1214.9 - \frac{44.2 \times 244}{5} = -942.06$$

and

$$r = \frac{S_{uv}}{\sqrt{S_{uu}S_{vv}}} = \frac{-942.06}{\sqrt{204.432 \times 4393.3}} = -0.994 \quad \text{(to 3 d.p.)}$$

In this case, the coding of the data points has **not** changed the value of the correlation coefficient.

Similarly, if you have worked through question 14 of Exercise 1 then you will have seen that if data values (x_i, y_i) are coded to obtain new data values (u_i, v_i) using the rules $u_i = 2x_i - 1$ and $v_i = 0.3y_i + 2$ then the correlation coefficient for the new data values was exactly the same as the correlation coefficient for the original data values.

In general

> If data values (x_i, y_i) are transformed to (u_i, v_i) by linear rules
>
> $u_i = ax_i + b \qquad v_i = cy_i + d$
>
> where
>
> $a > 0$ and $c > 0$
>
> then
>
> the correlation coefficient of u and v is the **same** as the correlation coefficient of x and y

A proof of this result is presented in Extension 1 at the end of the chapter.

Regression Lines

Once we have drawn a scatter diagram and calculated the correlation coefficient, you may feel that it is appropriate to try to fit a straight line to the data.

When you have fitted straight lines to data in the past, you would have done so by eye, which is rather haphazard and lacks consistency. It is possible to establish a formal procedure for calculating the best fit line.

Consider the data given in the table and its associated scatter graph.

x	y
2.6	9.8
3.2	8.1
4.1	7.2
5.9	6.9
7.1	4.3

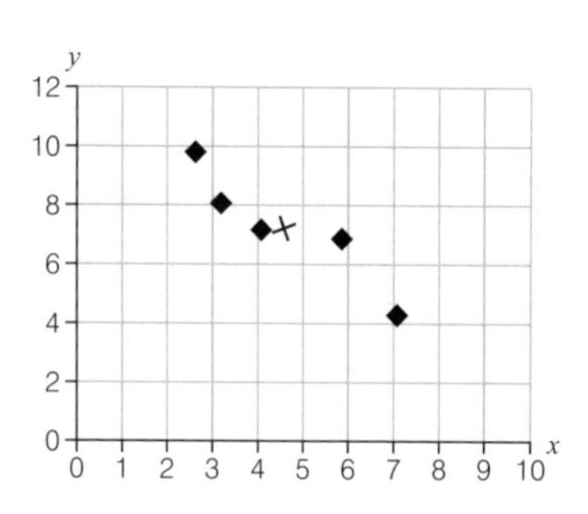

You can see that the data points are reasonably closely packed about a line of negative gradient.

If the correlation coefficient is calculated on a calculator, the value

$$r = -0.935 \qquad \text{(3 d.p.)}$$

is obtained and this confirms the fact that there should be a line of negative gradient which approximates the data points reasonably.

In the diagram, four lines have been drawn onto the scatter diagram:

A: $y = -x + 14$
B: $y = -x + 12$
C: $y = -1.5x + 14$
D: $y = -1.5x + 12$

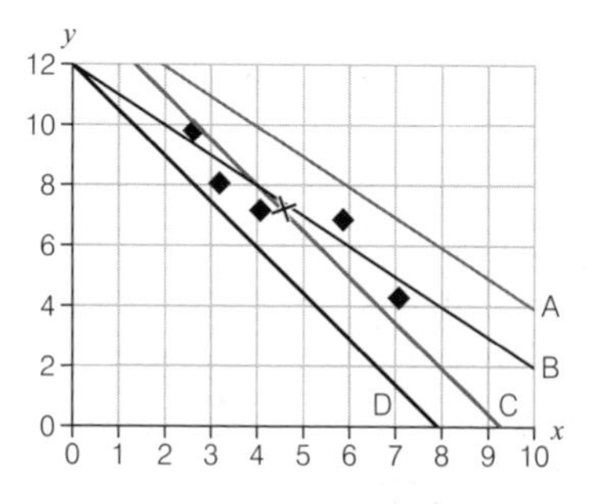

Clearly lines A and D would be very poor attempts at a best fit line. On the other hand lines B and C would be acceptable attempts at a drawn-by-eye line to fit the data.

Measuring the Error in Fitting a Line to the Data Points

Consider the line

$$y = mx + c$$

as a fit for the data points.

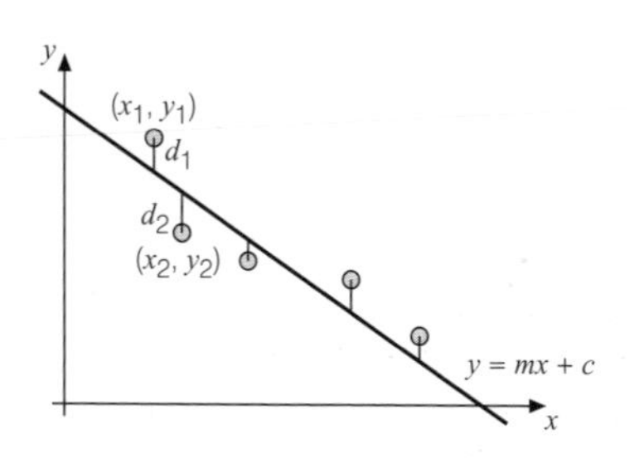

Let the **vertical difference** between the data point (x_i, y_i) and the line be d_i.

The point on the line with x co-ordinate x_i is

$$(x_i, mx_i + c)$$

Then

$$d_i = y_i - (mx_i + c)$$

and d_i will be positive if the data point is above the line and negative if the data point is below the line.

The error involved in using $y = mx + c$ for the data points could be taken as $\Sigma|d_i|$ or $\Sigma\, d_i^2$. As you will have noticed before, it is easier to work with expressions involving squares than expressions involving moduli.

The standard way of measuring the error in using the line $y = mx + c$ to estimate the data points is therefore to calculate the sum of the squares of these vertical distances. If you write E for this error, we have

$$E = \sum d_i^2 = \sum (y_i - mx_i + c)^2$$

The error for each of the four lines A, B, C and D can be calculated quickly using a spreadsheet.

A: y = −x + 14

x	y	y line	d	d^2
2.6	9.8	11.4	−1.6	2.56
3.2	8.1	10.8	−2.7	7.29
4.1	7.2	9.9	−2.7	7.29
5.9	6.9	8.1	−1.2	1.44
7.1	4.3	6.9	−2.6	6.76
			$\sum d^2 =$	25.34

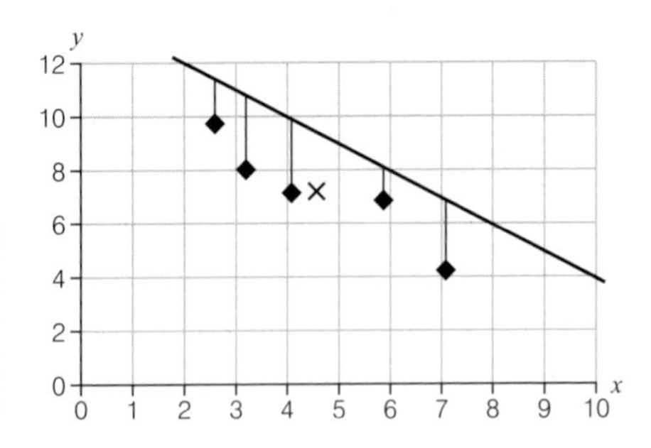

Error for line A = 25.34

B: y = −x + 12

x	y	y line	d	d^2
2.6	9.8	9.4	0.4	0.16
3.2	8.1	8.8	−0.7	0.49
4.1	7.2	7.9	−0.7	0.49
5.9	6.9	6.1	0.8	0.64
7.1	4.3	4.9	−0.6	0.36
			$\sum d^2 =$	2.14

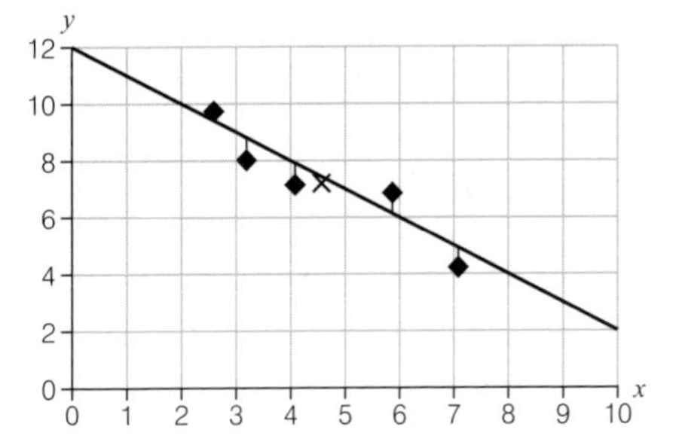

Error for line B = 2.14

C: y = −1.5x + 14

x	y	y line	d	d^2
2.6	9.8	10.1	−0.3	0.09
3.2	8.1	9.2	−1.1	1.21
4.1	7.2	7.85	−0.65	0.4225
5.9	6.9	5.15	1.75	3.0625
7.1	4.3	3.35	0.95	0.9025
			$\sum d^2 =$	5.6875

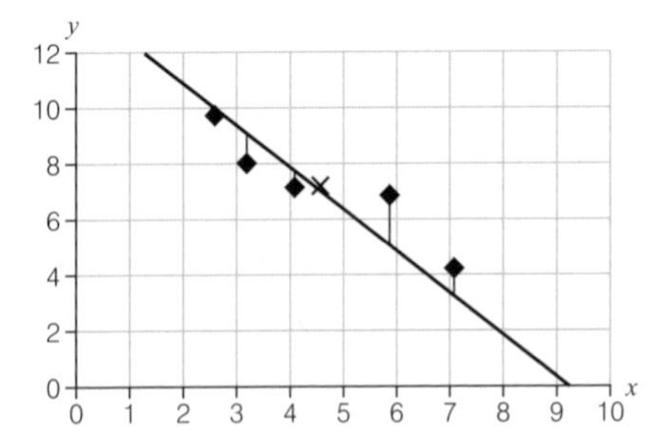

Error for line C = 5.6875

D: $y = -1.5x + 12$

x	y	y line	d	d^2
2.6	9.8	8.1	1.7	2.89
3.2	8.1	7.2	0.9	0.81
4.1	7.2	5.85	1.35	1.8225
5.9	6.9	3.15	3.75	14.0625
7.1	4.3	1.35	2.95	8.7025
			$\sum d^2 =$	28.2875

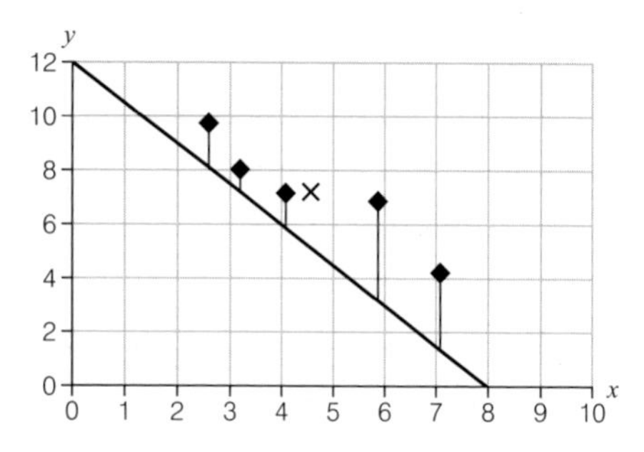

Error for line D = 28.2875

From these calculations, you can see that line B is the best of the four lines to the extent that it has the smallest sum of squares of the vertical distances.

However, the equation of the line which gives the smallest possible value for the error, E, has not yet been found.

The y on x Regression Line

It can be shown that the line that minimises the sum of the squares of the **vertical** distances between the line and the data points has equation

$$y - \bar{y} = \frac{S_{xy}}{S_{xx}}(x - \bar{x})$$

This line is called the **y on x regression line**.

If you recall from module C1 that the line through (a, b) with gradient m has equation $y - b = m(x - a)$, you can deduce that the y on x regression line:

- passes through the point $(\bar{x}, \bar{y})$;
- has gradient $\frac{S_{xy}}{S_{xx}}$: this number is called the **y on x coefficient of regression**.

A derivation of the equation of the y on x regression line is presented as Extension 2 at the end of this chapter.

The y on x line

- minimises the sum of the squares of the **vertical** distances between the line and the data points
- has equation

$$y - \bar{y} = \frac{S_{xy}}{S_{xx}}(x - \bar{x})$$

- the number $\frac{S_{xy}}{S_{xx}}$ is called the y on x coefficient of regression.

You can now calculate the equation of the y on x regression line for the example:

x	y	x^2	xy
2.6	9.8	6.76	25.48
3.2	8.1	10.24	25.92
4.1	7.2	16.81	29.52
5.9	6.9	34.81	40.71
7.1	4.3	50.41	30.53
22.9	36.3	119.03	152.16

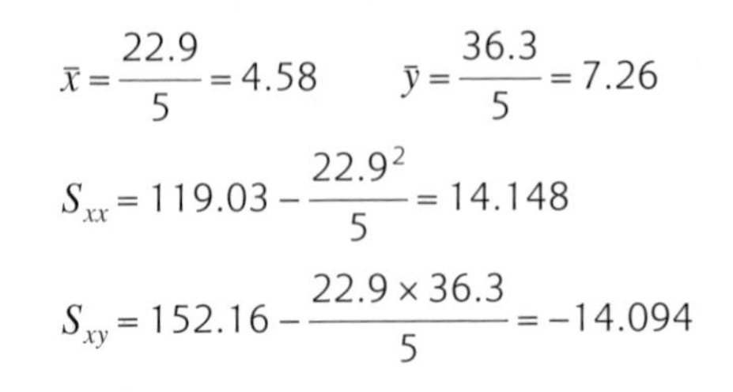

$$\bar{x} = \frac{22.9}{5} = 4.58 \qquad \bar{y} = \frac{36.3}{5} = 7.26$$

$$S_{xx} = 119.03 - \frac{22.9^2}{5} = 14.148$$

$$S_{xy} = 152.16 - \frac{22.9 \times 36.3}{5} = -14.094$$

so the y *on* x regression line

$$y - \bar{y} = \frac{S_{xy}}{S_{xx}}(x - \bar{x})$$

is

$$y - 7.26 = \frac{-14.094}{14.148}(x - 4.58)$$

or

$$y = -0.9962x + 11.82$$

Your calculator should be able to work out the equation of the y on x regression line. Make sure you know how!

where the coefficients are correct to 4 s.f.

It can easily be verified that this line is indeed better than any of the lines previously considered.

x	y	y **line**	d	d^2
2.6	9.8	9.22988	0.57012	0.325037
3.2	8.1	8.63216	−0.53216	0.283194
4.1	7.2	7.73558	−0.53558	0.286846
5.9	6.9	5.94242	0.95758	0.916959
7.1	4.3	4.74698	−0.44698	0.199791
			$\sum d^2 =$	2.011828

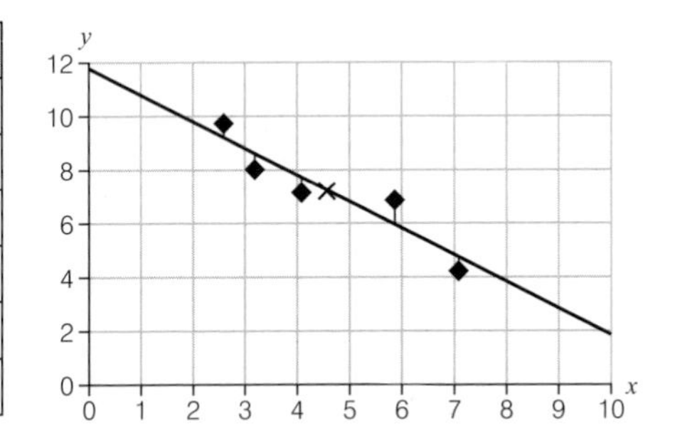

The measure of error for this line is 2.012 (3 d.p.), which is less than the error of each of the earlier lines.

Using the y on x Regression Line

Once the y on x regression line has been calculated it can be used to predict the y values we might have corresponding to new x values.

EXAMPLE 1

The table below shows the highest (x) and lowest (y) values (in pence) of the shares of 10 electronics companies for the 12-month period May 2003–April 2004.

High (x)	355	63	202	368	40	329	171	247	95	287
Low (y)	200	31	143	211	14	167	102	132	37	169

Find the equation of the y on x regression line and hence estimate the lowest value over this 12-month period of another electronics company share which had a high of 320p.

SOLUTION

High (x)	Low (y)	x^2	xy
355	200	126 025	71 000
63	31	3969	1953
202	143	40 804	28 886
368	211	135 424	77 648
40	14	1600	560
329	167	108 241	54 943
171	102	29 241	17 442
247	132	61 009	32 604
95	37	9025	3515
287	169	82 369	48 503
2157	1206	597 707	337 054

$$\bar{x} = \frac{2157}{10} = 215.7 \qquad \bar{y} = \frac{1206}{10} = 120.6$$

$$S_{xx} = 597\,707 - \frac{2157^2}{10} = 132\,442.1$$

$$S_{xy} = 337\,054 - \frac{2157 \times 1206}{10} = 76\,919.8$$

so the y on x regression line

$$y - \bar{y} = \frac{S_{xy}}{S_{xx}}(x - \bar{x})$$

is

$$y - 120.6 = \frac{76\,919.8}{132\,442.1}(x - 215.7)$$

or

$$y = 0.5808x - 4.674$$

where the constants are correct to 4 s.f.

If the regression line is to be used to make predicted values correct to three significant figures, it is worth giving the values of the constants in the equation of the regression line correct to 4 or even 5 significant figures.

Putting $x = 320$ into this equation gives

$$y = 0.5808 \times 320 - 4.674 = 181\text{p (to the nearest penny).}$$

The equation $y = 0.5808x - 4.674$ describes how the lowest price of the share, y, depends on the highest price of the share, x. The lowest price of the share, y, is called the **dependent** variable and the highest price of the share, x, is called the **independent** variable.

The y on x regression line allows us to predict a value for the dependent variable, y, from a known value of the independent variable, x.

The Reliability of Predictions made from a Regression Line

You have seen that you can use the y on x regression line to predict the y value of an observation if you know just the x value. This can often be useful: for example, x and y may be two properties of a material, x being very easily measured but y being very difficult or expensive to measure.

The reliability of such a prediction depends on being sure that a straight line fits the data points well and that the straight line fit is still applicable for the value of the independent variable, x, under consideration.

Interpolation is when the x value is within the range of already observed x values. Provided the straight line fit is good (i.e. the value of the correlation coefficient is close to 1 or −1) then interpolation is likely to provide a reliable prediction of the y value.

Extrapolation is when the x value is outside the range of already observed x values.

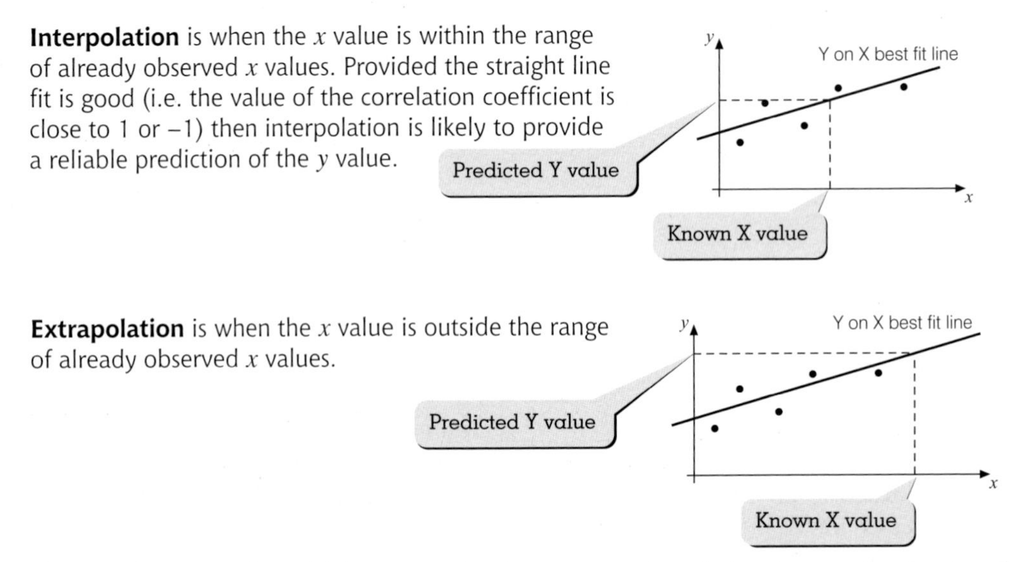

If extrapolation is to be reliable you need to be sure that the calculated regression line is still valid outside the interval of observed x values. Very rarely will we be certain that this is the case. (For example, you may have done experiments where weights, x Newtons, are hung on an elastic string and the length, y cm, of the spring is measured. The relationship between y and x is linear up to a certain point (called the elastic limit) but beyond that point the linear relationship fails completely.)

As a general rule, predictions obtained through extrapolation should be viewed cautiously since they may well be unreliable. The further the x value is away from the observed data points, the less trustworthy the prediction!

> Predictions using the y on x regression line of values of the dependent variable y from a given value of the independent variable x may be regarded as reliable provided
>
> the value of r for the data points is close to 1 or −1
>
> and
>
> the prediction process is using interpolation (i.e. the given x value is within the range of x values of the data points)

EXERCISE 2

1 A survey was conducted at an office into the value of a person's housing and their commuting time.

Value, x (thousands of £)	128	166	182	90	334	196	332	480	358	420
Daily commuting time, y (minutes)	160	140	150	210	90	110	100	50	40	70

a) Calculate the equation of the y on x regression line for this data.
b) Estimate the commuting time of a person whose house value is:
i) £280 000; **ii)** £650 000.
c) Given that the product moment correlation coefficient for this data is −0.916, correct to three decimal places, comment on the reliability of the estimates made in (b).

2 The table shows the gestation period (x days) and the maximum recorded life span (y years) of different species of animal:

Animal	chimp	dog	hedgehog	hippo	horse	lion	panda	pig	rabbit	squirrel
Gestation period x days	235	60	30	240	350	110	138	115	30	40
Maximum life span y years	56	29	16	54	62	29	27	27	18	23

a) Calculate the equation of the y on x regression line for this data.
b) Use this line to estimate the maximum life span of a species of animal whose gestation period is 90 days. Comment on the reliability of your estimate. (The correlation coefficient for this data is 0.957.)

3 The table below shows the age (x years) and the mass (y lb) of 10 randomly chosen primary school children:

x	7.8	6.2	7.0	6.4	6.0	8.0	5.6	8.0	5.2	7.3
y	61	48	52	56	50	60	44	57	38	57

Construct a scatter diagram for the data and determine the least squares regression line of y on x. Use the line to predict the mass of a primary school child whose age is 7.0 years.

Explain why it might be unwise to use the same relationship to estimate the mass of a primary school child of age 10.5 years.

4 A test was performed to determine the relationship between the chemical content (y g/l) of a particular constituent in solution and the crystallisation temperature (x °C). The crystallisation temperatures observed in the tests were in the range 0 °C to 7 °C. The results can be summarised as

$n = 7$
$\sum x = 16.8$ $\sum x^2 = 62.32$ $\sum y = 38.5$ $\sum y^2 = 245.49$ $\sum xy = 119.36$

a) Calculate the product moment correlation coefficient for the data.
b) Find the equation of the y on x regression line for this data.
c) Use this line to predict the chemical content of the constituent in the cases when the crystallisation temperature is:
i) 5 °C; **ii)** 12 °C.
d) Comment on the reliability of these predictions.

5 The table below shows the heights and ages of a randomly chosen sample of a species of fir trees:

Age (x years)	0.5	1.0	1.5	2.0	2.5	3.0	3.5	4.0
Height (y m)	0.05	0.25	0.29	0.54	0.71	0.76	0.98	1.05

a) Draw a scatter diagram for the data.
b) Calculate the equation of the y on x regression line. Draw the regression line on your scatter diagram.
c) Estimate the height of a 21-month old fir tree.

The x on y Regression Line

The y on x line minimised the sum of the squares of the **vertical** distances of the data points from the line.

In a similar way, the ***x* on *y* line** minimises the sum of the squares of the **horizontal** distances of the data points from the line.

Notice that when working with the x on y line it is usual to write the line equation in the $x = ay + b$ format.

The error in using the line $x = ay + b$ as a fit for the data points is measured by $\sum h_i^2$ where $h_i = x_i - (ay_i + b)$

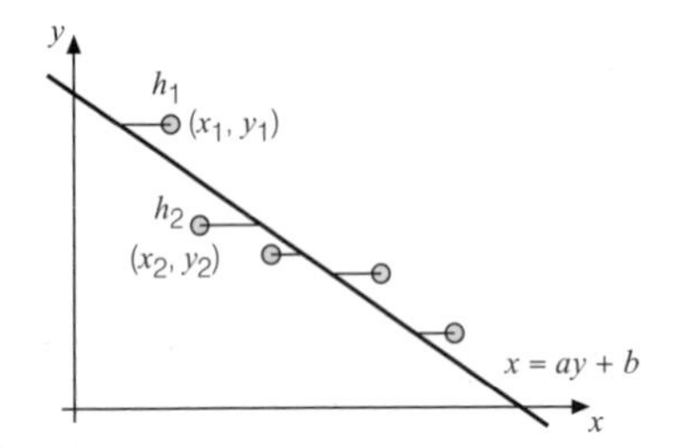

Using methods very similar to the ones used in the derivation of the equation of the y on x regression line (see Extension 2) it can be shown that

The x on y line

- minimises the sum of the squares of the **horizontal** distances between the line and the data points
- has equation

$$x - \bar{x} = \frac{S_{xy}}{S_{yy}}(y - \bar{y})$$

- the number $\frac{S_{xy}}{S_{yy}}$ is called the x on y coefficient of regression.

The details of the calculation of the x on y line are very similar to those of the y on x line. For the data set

x	y
2.6	9.8
3.2	8.1
4.1	7.2
5.9	6.9
7.1	4.3

you need to calculate $\bar{x}$, $\bar{y}$, $\sum y^2$ and $\sum xy$:

x	y	y^2	xy
2.6	9.8	96.04	25.48
3.2	8.1	65.61	25.92
4.1	7.2	51.84	29.52
5.9	6.9	47.61	40.71
7.1	4.3	18.49	30.53
22.9	36.3	279.59	152.16

$$\bar{x} = \frac{22.9}{5} = 4.58,\ \bar{y} = \frac{36.3}{5} = 7.26$$

$$S_{yy} = 279.59 - \frac{36.3^2}{5} = 16.052$$

$$S_{xy} = 152.16 - \frac{22.9 \times 36.3}{5} = -14.094$$

so the x on y regression line

$$x - \bar{x} = \frac{S_{xy}}{S_{yy}}(y - \bar{y})$$

is

$$x - 4.58 = \frac{-14.094}{16.052}(y - 7.26)$$

or

$$x = -0.8780y + 10.95$$

where the coefficients are correct to 4 s.f.

Calculators usually work out the equation of the y on x line. If you wish to evaluate the equation of the x on y line it is probably best to use the calculator just to evaluate $\bar{x}$, $\bar{y}$, $\sum y^2$ and $\sum xy$ and then use the method shown here.

This x on y regression line can be added easily to a scatter diagram: it is known to pass through the point $(\bar{x}, \bar{y})$ and the equation $x = -0.8780y + 10.95$ implies that when $y = 0$, $x = 10.95$.

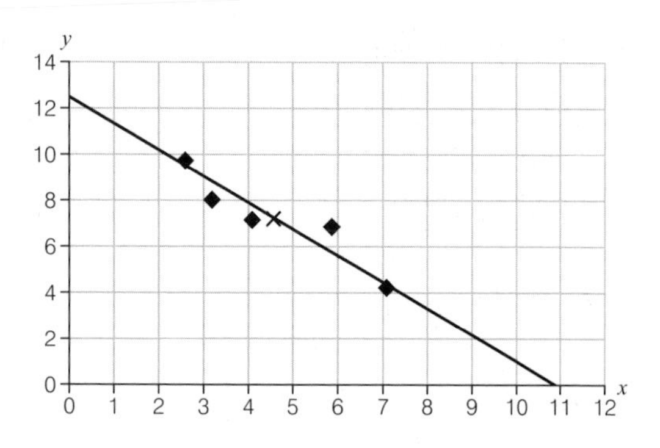

EXERCISE 3

1 Draw a scatter diagram on graph paper for the data shown in the table.

x	3.2	5	7.4	8.4
y	0.7	6.1	6.2	7

a) Calculate the product moment correlation coefficient for the data.
b) Calculate the equation of the x on y regression line for this data. Draw the line on your diagram.
c) The y on x line has equation $y = 1.0519x - 1.3116$, where the coefficients are correct to four decimal places. Draw this line on your diagram.
d) Write down, or calculate, the co-ordinates of the point of intersection of the two lines.

2 The table shows the gestation period (x days) and the maximum recorded life span (y years) of different species of animal.

Animal	chimp	dog	hedgehog	hippo	horse	lion	panda	pig	rabbit	squirrel
Gestation period x days	235	60	30	240	350	110	138	115	30	40
Maximum life span y years	56	29	16	54	62	29	27	27	18	23

a) Calculate the equation of the x on y regression line for this data.
b) Use this line to estimate the gestation period of a species of animal whose maximum life span is 50 years. Comment on the reliability of your estimate. (The correlation coefficient for this data is 0.957.)

The Relationship Between the Two Regression Lines

It is known that the y on x line has equation $y - \bar{y} = \frac{S_{xy}}{S_{xx}}(x - \bar{x})$.

The x on y line has equation $x - \bar{x} = \frac{S_{xy}}{S_{yy}}(y - \bar{y})$, which can be rewritten as $y - \bar{y} = \frac{S_{yy}}{S_{xy}}(x - \bar{x})$.

Notice that both lines pass through the point $(\bar{x}, \bar{y})$. This means that the point $(\bar{x}, \bar{y})$ will always be the point of intersection of the two lines.

If the correlation coefficient, r, is very close to 1 or –1 then r^2 will be very close to 1:

$$\Rightarrow \quad \frac{S_{xy}^2}{S_{xx}S_{yy}} \approx 1$$

$$\Rightarrow \quad \frac{S_{xy}}{S_{xx}} \approx \frac{S_{yy}}{S_{xy}}$$

so the y on x line

$$y - \bar{y} = \frac{S_{xy}}{S_{xx}}(x - \bar{x})$$

will be very close to the x on y line

$$y - \bar{y} = \frac{S_{yy}}{S_{xy}}(x - \bar{x})$$

If r is very close to 1 or −1, the y on x and x on y lines will be very close to each other.

Which Regression Line Should be Used to Make Predictions?

So far, in all the examples that have been considered, both the x and y values have been free to vary randomly. When this is the case, you should use the y on x line to estimate a y value from a given x value since the y on x line minimises the vertical errors and should therefore give you the estimate with the least possible error. Similarly, when x and y values are free to vary randomly then you should use the x on y line to estimate an x value from a given y value. The x on y line minimises the horizontal errors and should give an x estimate with the least possible error.

Sometimes, however, the person conducting the survey will be able to choose the values for one of the variables (usually, but not necessarily, the x values). This variable is not free to vary randomly since it has been controlled. Such a variable is called a **control variable**.

When the x values are controlled there is no error in the x values – all the error is due to the variability of the y values. You should therefore use the y on x line, which minimises the vertical errors, for making all predictions from such data. Similarly, if the y values are controlled then there is no error in the y values and all the error is due to the variability of the x values. The x on y line, which minimises the horizontal errors, should therefore be used for making all predictions from such data.

If both x and y values can vary randomly

- use y on x line to show how y depends on x and to predict a y value from a given x value
- use x on y line to show how x depends on y and to predict x values from a given y value

If x is a controlled variable

- use the y on x line for **all** predictions

If y is a controlled variable

- use the x on y line for **all** predictions

However, if the value of r is very close to 1 or −1 then there is very little difference between the two lines so reliable predictions can be obtained from either line.

EXAMPLE 2

The table below shows the marks of 10 students from two short tests: one for memory (X) and one for intelligence (Y).

Memory	5	8	7	10	4	7	9	6	8	6
Intelligence	7	9	6	9	6	7	10	7	6	8

EXAMPLE 2 (continued)

a) Draw a scatter diagram and calculate the correlation coefficient for the data.

b) Determine the equations of the y on x and the x on y regression lines for the data and show each line on your diagram.

c) Use the appropriate line to determine:

  i) an estimate for the score on the intelligence test for a student who scored 8.5 on the memory test;

  ii) an estimate for the score on the memory test for a student who scored 7 on the intelligence test.

SOLUTION

In this example both x and y are free to vary randomly: the teacher who recorded these results has certainly not controlled either test mark.

a) The calculator gives

$$n = 10$$

$$\Sigma x = 70 \qquad \Sigma x^2 = 520 \qquad \Sigma y = 75 \qquad \Sigma y^2 = 581 \qquad \Sigma xy = 540$$

so

$$\bar{x} = \frac{\Sigma x}{n} = \frac{70}{10} = 7 \qquad \bar{y} = \frac{\Sigma y}{n} = \frac{75}{10} = 7.5$$

$$S_{xx} = \Sigma x^2 - \frac{(\Sigma x)^2}{n} = 520 - \frac{70^2}{10} = 30 \qquad S_{yy} = \Sigma y^2 - \frac{(\Sigma x)^2}{n} = 581 - \frac{75^2}{10} = 18.5$$

$$S_{xy} = \Sigma xy - \frac{\Sigma x \Sigma y}{n} = 540 - \frac{70 \times 75}{10} = 15$$

$$r = \frac{S_{xy}}{\sqrt{S_{xx} S_{yy}}} = \frac{1.5}{\sqrt{30 \times 18.5}} = 0.637 \quad \text{(3 d.p.)}$$

b) The y on x regression line is

$$y - \bar{y} = \frac{S_{xy}}{S_{xx}}(x - \bar{x}) \quad \Rightarrow \quad y - 7.5 = \frac{15}{30}(x - 7) \quad \Rightarrow \quad y = 0.5x + 4$$

The x on y regression line is

$$x - \bar{x} = \frac{S_{xy}}{S_{yy}}(y - \bar{y}) \quad \Rightarrow \quad x - 7 = \frac{1.5}{1.85}(y - 7.5) \quad \Rightarrow \quad x = \frac{30}{37}y + \frac{34}{37}$$

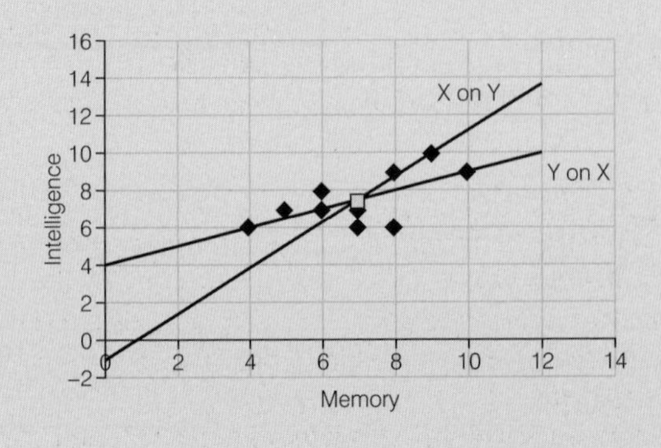

EXAMPLE 2 (continued)

c) To estimate y given a value of x you must use the $\boldsymbol{y}$ **on** $\boldsymbol{x}$ line: $y = 0.5x + 4$
when $x = 8.5$, $y = 0.5 \times 8.5 + 4 = 8.25$
To estimate x given a value of y you must use the $\boldsymbol{x}$ **on** $\boldsymbol{y}$ line: $x = \frac{30}{37}y + \frac{34}{37}$
when $y = 7$, $x = \frac{30}{37} \times 7 + \frac{34}{37} = 6.59$ (3 s.f.)

EXAMPLE 3

A forestry worker measured the height of a sapling when it was planted and at intervals of six months. She gathered the following data:

Time t (months)	0	6	12	18	24	30
Height h (m)	0.45	0.78	0.97	1.31	1.57	1.83

a) Calculate the product moment correlation coefficient for the data.
b) By using appropriate regression lines, determine:
   i) an estimate for the height of the sapling after 10 months growth;
   ii) the time when the height of the sapling reached 1.5 m.
c) Comment on the reliability of your answers to (b).

SOLUTION

a) Inputting the data on a calculator you will obtain $r = 0.998$ (3 d.p.)
b) Since the t data is certainly controlled by the forestry worker, the h on t line should be used for all predictions.

A calculator gives the equation of the h on t line as $h = 0.04576t + 0.4652$ (coefficients given to 4 s.f.)
i) Putting $t = 10$ gives $h = 0.04576 \times 10 + 0.4652 = 0.923$ m (3 s.f.)
ii) Putting $h = 1.5$ gives $1.5 = 0.04576t + 0.4652$
$\Rightarrow 1.0348 = 0.04576t$
$\Rightarrow t = 22.6$ months (3 s.f.)

c) Since $t = 10$ and $h = 1.5$ are inside the ranges of h and t values of the data points, we are interpolating rather than extrapolating. This, together with the fact that r is very close to 1, makes these estimates reliable.

EXERCISE 4

1 A liquid which is stored in 10 litre drums is subject to evaporation loss. It is believed that the quantity of liquid remaining in a drum which has been stored for five days or longer is a linear function of the duration of storage. A sample of five drums, each initially containing 10 litres of the liquid, was stored under identical conditions and on every second day from the sixth to the fourteenth, one drum was opened and the quantity of liquid in it was determined. The results obtained were:

Duration of storage (x days)	6	8	10	12	14
Quantity of liquid (y litres)	9.6	9.3	9.1	8.9	8.6

a) Draw a scatter diagram for the data and use a suitable statistic to determine whether the assumption of a linear relationship between amount remaining and storage duration is reasonable.
b) Calculate the equation of an appropriate regression line for the data.

c) Use your answers to part (b) to determine estimates for:

i) the amount of liquid remaining after 9 days of storage;
ii) the duration of storage required for the amount of liquid to come down to 8.8 litres.
iii) the amount of liquid remaining after 25 days of storage.

d) Comment on the reliability, or otherwise, of your answers to part (c).

2 A scientist, working in an agricultural research station, believes that there is a linear relationship between the hardness of the shells of eggs laid by chickens (y which is measured on a 0–10 scale with 10 being the hardest) and the amount of a certain food supplement (x grams per week) put into the diet of the chickens. He selects 10 chickens of the same breed and collects the following data:

x	7.0	9.8	11.6	17.5	7.6	8.2	12.4	17.5	9.5	19.5
y	1.2	2.1	3.4	6.1	1.3	1.7	3.4	6.2	2.1	7.1

a) Plot the data on a scatter diagram.
b) Calculate the equations of the y on x and x on y regression lines.
c) Determine an estimate of the hardness of shell that should be obtained from a chicken who receives 10 grams per week of the food supplement.
d) Determine an estimate of the level of food supplement necessary to produce an egg hardness of 5.

3 When loads of size x Newtons were attached to an elastic string and the extended lengths, y cm, of the string were measured the following results were obtained:

Load (x N)	5	8	12	17	23
Length (y cm)	5.8	7.4	8.5	9.7	11.3

$\Sigma x = 65$, $\Sigma y = 42.7$, $\Sigma x^2 = 1051$, $\Sigma y^2 = 382.43$, $\Sigma xy = 615$

a) Calculate the product moment correlation coefficient for the data.
b) By first finding the equation of an appropriate regression line, determine the load that must be attached to the string if the string is to have a length of 9 cm.

4 A supermarket conducted a survey to investigate how the time, t seconds, it took to process a customer at the checkout varied with the number, n, of items the customer was buying. The results they obtained were:

Time (t s)	74	97	113	192	210	190	338
n	4	8	12	16	20	30	40

$\Sigma t = 1214$, $\Sigma n = 130$, $\Sigma t^2 = 258\,962$, $\Sigma n^2 = 3380$, $\Sigma tn = 28\,920$

a) Calculate the product moment correlation coefficient for the data.
b) By first finding the equation of an appropriate regression line, estimate the time required to process a customer who is buying 25 items.

5 The price £x of a digital camera is reduced by £30 every three months. The number of cameras sold during the period before the next decrease is y thousand. The values covering six consecutive periods are shown in the table:

x	450	420	390	360	330	300
y	8.5	9.8	10.9	12.6	13.2	14.2

$\Sigma x = 2250$, $\Sigma x^2 = 859\ 500$, $\Sigma y = 69.2$, $\Sigma y^2 = 821.74$, $\Sigma xy = 25\ 344$

a) Plot a scatter diagram for the data.

b) Obtain the equation of an appropriate regression line for the data and plot this line on your scatter diagram.

c) Calculate an estimate of the number of cameras that will be sold when the price is £240. Comment on the reliability of your estimate.

6 When I travel to work each morning, my arrival time depends on when I depart. Over a period of 10 consecutive days I made a note of my departure and arrival times. In the following table, x is the departure time, in minutes after 7 a.m., and y is the arrival time, in minutes after 8 a.m.

x	20	27	33	40	45	50	53	58	65	70
y	4	8	22	30	44	56	70	80	85	88

$\Sigma\ 461$, $\Sigma y = 487$, $\Sigma x^2 = 23\ 641$, $\Sigma y^2 = 32\ 805$, $\Sigma xy = 27\ 037$

i) Plot a scatter diagram to illustrate the data.

ii) Calculate the product moment correlation coefficient, and explain what information your answer gives you about the relationship between x and y.

iii) Find the equation of the regression line of y on x in the form $y = a + bx$, giving a and b correct to 3 significant figures.

iv) I wish to estimate the time at which I should depart in order to arrive at 8.50 a.m. Give a reason why it will make little difference whether the regression line of y on x or the regression line of x on y is used.

Use the equation obtained in part (iii) to calculate the required estimate.

v) Give a reason why it would probably not be sensible to use either line to estimate the value of y when $x = 120$.

(OCR Nov 96 S1)

EXTENSION 1

Understanding the Formula and Properties of the Correlation Coefficient

The Covariance of a Set of Data Points

Looking back at the scatter diagram for data set 1, if the lines $x = \bar{x}$, $y = \bar{y}$ are drawn in then the diagram is divided into four quadrants:

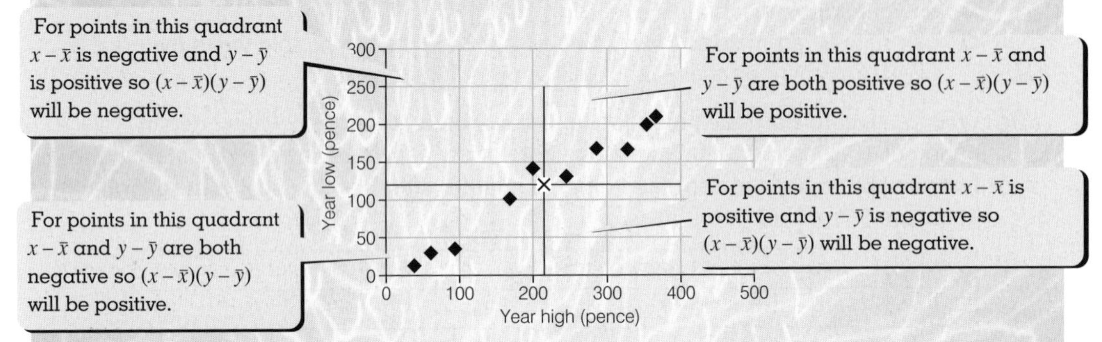

In this case, where the data points are closely packed around a line of positive gradient, you can see that most of the points lie in the top right and bottom left quadrants.

If you calculate $(x - \bar{x})(y - \bar{y})$ for each data point, you can see from the diagram that all but one of the answers will be positive. The average of these values will certainly be positive.

The average of the $(x - \bar{x})(y - \bar{y})$ values is called the **covariance** of the data. If there are n data points then

$$\text{Cov}(x, y) = \frac{1}{n} \Sigma (x - \bar{x})(y - \bar{y})$$

Notice that the units of covariance will be the units of x multiplied by the units of y.

Now consider data set 2:

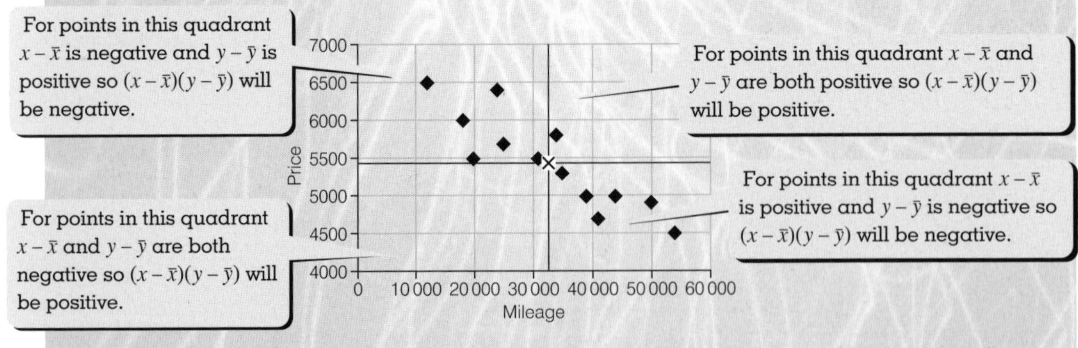

In this case, where the data points are closely packed around a line of negative gradient, you will see that most of the points lie in the top left and bottom right quadrants.

If you calculate $(x - \bar{x})(y - \bar{y})$ for each data point, you can see from the diagram that most of the answers will be negative. The covariance, or average of these values, will certainly be negative.

Now consider data set 3:

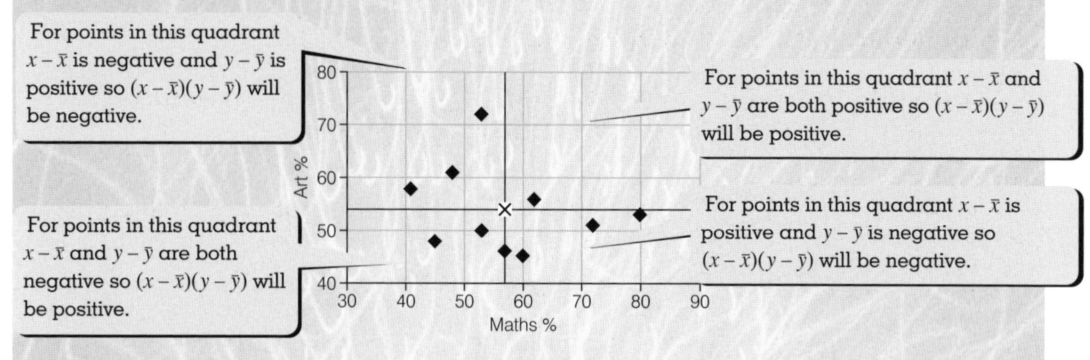

In this case, the data points are spread fairly evenly over all four quadrants.

If you calculate $(x - \bar{x})(y - \bar{y})$ for each data point, there will be an almost equal number of negative and positive values. The covariance, or average of these values, will be close to 0.

From these three examples you can see that

- data points clustered around a line of positive gradient will give a positive value of covariance
- data points clustered around a line of negative gradient will give a negative value of covariance
- data points which are scattered all over the place will give a covariance close to 0

Evaluation of the Covariance

The covariance of data points $(x_1, y_1), (x_2, y_2), \ldots, (x_n, y_n)$ can be calculated by first evaluating

$$S_{xy} = \Sigma\,(x_i - \bar{x})(y_i - \bar{y})$$

and then dividing the result by n.

So $\quad \text{Cov}(x, y) = \dfrac{S_{xy}}{n}$

Alternative Formulae for S_{xy}

If there are n data points $(x_1, y_1), (x_2, y_2), \ldots, (x_n, y_n)$ then

$$S_{xy} = \Sigma\,(x_i - \bar{x})(y_i - \bar{y}) = \Sigma\,x_i y_i - n\bar{x}\bar{y} = \Sigma\,x_i y_i - \frac{(\Sigma\,x_i)(\Sigma\,y_i)}{n}$$

Proof:

$$
\begin{aligned}
S_{xy} &= \Sigma\,(x_i - \bar{x})(y_i - \bar{y}) \\
&= (x_1 - \bar{x})(y_1 - \bar{y}) + (x_2 - \bar{x})(y_2 - \bar{y}) + \cdots + (x_n - \bar{x})(y_n - \bar{y}) && [1] \\
&= x_1 y_1 - x_1\bar{y} - \bar{x}y_1 + \bar{x}\bar{y} + x_2 y_2 - x_2\bar{y} - \bar{x}y_2 + \bar{x}\bar{y} + \cdots + x_n y_n - x_n\bar{y} - \bar{x}y_n + \bar{x}\bar{y} \\
&= (x_1 y_1 + x_2 y_2 + \cdots + x_n y_n) - (x_1\bar{y} + x_2\bar{y} + \cdots - x_n\bar{y}) - (\bar{x}y_1 + \bar{x}y_2 + \cdots + \bar{x}y_n) \\
&\quad + (\bar{x}\bar{y} + \bar{x}\bar{y} + \cdots + \bar{x}\bar{y}) \\
&= \Sigma\, x_i y_i - \bar{y}(x_1 + x_2 + \cdots - x_n) - \bar{x}(y_1 + y_2 + \cdots + y_n) + n\bar{x}\bar{y} \\
&= \Sigma\, x_i y_i - \bar{y}\,\Sigma\, x_i - \bar{x}\,\Sigma\, y_i + n\bar{x}\bar{y} \\
&= \Sigma\, x_i y_i - \bar{y}n\bar{x} - \bar{x}n\bar{y} + n\bar{x}\bar{y} \\
&= \Sigma\, x_i y_i - n\bar{x}\bar{y} && [2] \\
&= \Sigma\, x_i y_i - n\,\frac{(\Sigma\, x_i)}{n}\,\frac{(\Sigma\, y_i)}{n} \\
&= \Sigma\, x_i y_i - \frac{(\Sigma\, x_i)\,(\Sigma\, y_i)}{n} && [3]
\end{aligned}
$$

Lines [1], [2] and [3] are the required formulae.

The covariance can now easily be calculated using either a table of values to record the working or entering the data values directly onto a suitable graphical or scientific calculator.

High price ***x* pence**	355	63	202	368	40	329	171	247	95	287
Low price ***y* pence**	200	31	143	211	14	167	102	132	37	169

Consider for example the share prices of data list 1:

High (x)	**Low (y)**	xy
355	200	71 000
63	31	1953
202	143	28 886
368	211	77 648
40	14	560
329	167	54 943
171	102	17 442
247	132	32 604
95	37	3515
287	169	48 503
2157	1206	337 054

$n = 10$

$$S_{xy} = \Sigma\, x_i y_i - \frac{(\Sigma\, x_i)(\Sigma\, y_i)}{n}$$

$$= 337\,054 - \frac{2157 \times 1206}{10}$$

$$= 76\,919.8 \text{ pence}^2$$

$$\text{Cov}(x, y) = \frac{S_{xy}}{n} = \frac{76\,919.8}{10} = 7691.98 \text{ pence}^2$$

This exercise introduces some of the properties of covariance: it is recommended that you work through this exercise thoroughly before progressing to the algebra of the next section.

EXERCISE

1 The price and lifetime of 10 different brands of electric lightbulbs were measured:

Price (pence) x	12	18	27	22	42	61	53	39	53	35
Lifetime (hours) y	178	234	255	282	377	577	521	442	487	365

a) Draw a scatter diagram for the data.
b) Calculate the standard deviation of the prices, sd_x, and the standard deviation of the lifetimes, sd_y.
c) Calculate the covariance, $Cov(x, y)$, of this data.
d) Verify that, for this data, $Cov(x, y)$ is slightly smaller than $sd_x \times sd_y$.

2 A survey asked eight people to record the length of time they watched BBC1 and C4 during a given week:

Person	1	2	3	4	5	6	7	8
BBC1 (x hours)	12	7	9	15	8	9	18	10
C4 (y hours)	7	6	12	4	10	4	9	5

a) Draw a scatter diagram for the data.
b) Calculate the standard deviation of the BBC times, sd_x, and the standard deviation of the C4 times, sd_y.
c) Calculate the covariance, $Cov(x, y)$, of this data.
d) Verify that, for this data,

$$-sd_x \times sd_y < Cov(x, y) < sd_x \times sd_y$$

3 A survey was conducted at an office into the value of a person's housing and their commuting time:

Value, x (thousands of £)	128	166	182	90	334	196	332	480	358	420
Daily commuting time, y (minutes)	160	140	150	210	90	110	100	50	40	70

Calculate the covariance of this data.
a) Draw a scatter diagram for the data.
b) Calculate the standard deviation of the house values, sd_x, and the standard deviation of the travelling times, sd_y.
c) Calculate the covariance, $Cov(x, y)$, of this data.
d) Verify that, for this data,

$$-sd_x \times sd_y < Cov(x, y) < sd_x \times sd_y$$

4

x	1	3	5	7	9
y	0.8	6.2	11.6	17.0	22.4

Draw a scatter diagram for the data and verify that $\text{Cov}(x, y) = \text{sd}_x \times \text{sd}_y$.

Properties of Covariance

The covariance of a set of data points has several important properties.

Property 1: For any set of data, $-\text{sd}_x \times \text{sd}_y \leqslant \text{Cov}(x, y) \leqslant \text{sd}_x \times \text{sd}_y$

Proof: Let $u_i = \dfrac{x - \bar{x}}{(\text{sd})_x}$ and $v_i = \dfrac{y - \bar{y}}{(\text{sd})_y}$

then

$$(u_i - v_i)^2 \geqslant 0$$

A square number is always greater than or equal to 0.

$$\Rightarrow \quad u_i^2 - 2u_iv_i + v_i^2 \geqslant 0$$
$$\Rightarrow \quad u_i^2 + v_i^2 \geqslant 2u_iv_i$$

This gives n separate inequalities:

$$u_1^2 + v_1^2 \geqslant 2u_1v_1, \qquad u_2^2 + v_2^2 \geqslant 2u_2v_2, \qquad \ldots \qquad u_n^2 + v_n^2 \geqslant 2u_nv_n$$

and if all these inequalities are added up you will obtain

$$u_1^2 + v_1^2 + u_2^2 + v_2^2 + \cdots + u_n^2 + v_n^2 \geqslant 2u_1v_1 + 2u_2v_2 + \cdots + 2u_nv_n$$
$$\Rightarrow \quad (u_1^2 + u_2^2 + \cdots + u_n^2) + (v_1^2 + v_2^2 + \cdots + v_n^2) \geqslant 2u_1v_1 + 2u_2v_2 + \cdots + 2u_nv_n$$
$$\Rightarrow \quad \sum u_i^2 + \sum v_i^2 \geqslant 2\sum u_iv_i$$

but

Since $(\text{sd})_x^2 = \frac{1}{n}\sum (x - \bar{x})^2$

$$\sum u_i^2 = \sum \frac{(x_i - \bar{x})^2}{(\text{sd})_x^2} = \frac{1}{(\text{sd})_x^2}\sum (x_i - \bar{x})^2 = \frac{1}{(\text{sd})_x^2} \times n(\text{sd})_x^2 = n$$

and similarly

$$\sum v_i^2 = n$$

also

Since $\text{Cov}(x, y) = \frac{1}{n}\sum (x - \bar{x})(y - \bar{y})$

$$\sum u_iv_i = \sum \frac{(x - \bar{x})}{(\text{sd})_x}\frac{(y - \bar{y})}{(\text{sd})_y} = \frac{1}{(\text{sd})_x(\text{sd})_y}\sum (x - \bar{x})(y - \bar{y}) = \frac{1}{(\text{sd})_x(\text{sd})_y} \times n\,\text{Cov}(x, y)$$

These results mean that the inequality

$$\sum u_i^2 + \sum v_i^2 \geqslant 2\sum u_iv_i$$

becomes

$$2n \geqslant \frac{2n\,\text{Cov}(x, y)}{(\text{sd})_x(\text{sd})_y}$$
$$\Rightarrow \quad (\text{sd})_x(\text{sd})_y \geqslant \text{Cov}(x, y)$$

The second half of the inequality is proved in the same way but starting with the fact that

$$(u_i + v_i)^2 \geqslant 0$$

expanding and rearranging gives

$$u_i^2 + v_i^2 \geqslant -2u_i v_i$$

and then adding up the n inequalities leads to

$$\sum u_i^2 + \sum v_i^2 \geqslant -2\sum u_i v_i$$

which can be rewritten as

$$2n \geqslant -\frac{2n\,\text{Cov}(x, y)}{(\text{sd})_x(\text{sd})_y}$$

$$\Rightarrow \quad -(\text{sd})_x(\text{sd})_y \leqslant \text{Cov}(x, y)$$

Remember that multiplying each side of an inequality by a negative number reverses the direction of the inequality.

Property 2: If the data points all lie on a line of positive gradient then $\text{Cov}(x, y) = \text{sd}_x \times \text{sd}_y$ and if the data points all lie on a line of negative gradient then $\text{Cov}(x, y) = -\text{sd}_x \times \text{sd}_y$.

Proof: If the data points actually lie on a straight line whose equation is $y = ax + b$ then

$$\begin{aligned}\text{Cov}(x, y) &= \tfrac{1}{n}\sum (x_i - \bar{x})(y_i - \bar{y})\\ &= \tfrac{1}{n}\sum (x_i - \bar{x})(ax_i + b - \bar{y})\\ &= \tfrac{1}{n}\sum (x_i - \bar{x})(ax_i + b - (a\bar{x} + b))\\ &= \tfrac{1}{n}\sum (x_i - \bar{x})(ax_i - a\bar{x})\\ &= a \times \tfrac{1}{n}\sum (x_i - \bar{x})^2\\ &= a \times (\text{sd})_x^2\end{aligned}$$

Remember the coding result of Chapter 3: if data values $y_1, \ldots, y_n$ are linked to data values $x_1, \ldots, x_n$ by a linear rule $y_i = ax_i + b$ then

$$\bar{y} = a\bar{x} + b$$

and $(\text{sd})_y = a \times (\text{sd})_x$ if $a > 0$
or $\text{sd}_y = -a \times \text{sd}_x$ if $a < 0$

where sd_x is the standard deviation of the x values.

If the points lie on a line of **positive gradient** then $a > 0$ so $(\text{sd})_y = a \times (\text{sd})_x$ and you can write

$$\text{Cov}(x, y) = a \times (\text{sd})_x^2 = (\text{sd})_x \times a \times (\text{sd})_x = (\text{sd})_x(\text{sd})_y$$

On the other hand, if the points lie on a line of negative gradient then $a < 0$ and $\text{sd}_y = -a \times \text{sd}_x$ which means that you can write

$$\text{Cov}(x, y) = a \times (\text{sd})_x^2 = -(\text{sd})_x \times -a \times (\text{sd})_x = -(\text{sd})_x(\text{sd})_y$$

Note that it can also be proved that $\text{Cov}(x, y) = (\text{sd})_x(\text{sd})_y$ **only when** all the data points lie on a line of positive gradient and $\text{Cov}(x, y) = -(\text{sd})_x(\text{sd})_y$ **only when** all the data points lie on a line of negative gradient.

Moving to the Correlation Coefficient

The results given in the last section suggest that by dividing the covariance of the data by the product of the standard deviations, you will obtain a measure, on a scale running from −1 to +1, indicating how close a set of data is to a straight line.

Definition If the sample data consists of n pairs of measurements (x_i, y_i) then the **product moment coefficient of linear correlation** is given by

$$r = \frac{\text{Cov}(x, y)}{(\text{sd})_x(\text{sd})_y}$$

where $(\text{sd})_x$ is the standard deviation of the x values and $(\text{sd})_y$ is the standard deviation of the y values.

The following properties of r are immediate consequences of its definition and the properties of covariance which were established in the previous section.

Properties of *r*

- r has no units
- $-1 \leqslant r \leqslant 1$
  $r = 1$ if and only if all the points lie on a straight line of positive gradient
  $r = -1$ if and only if all the points lie on a straight line of negative gradient
- If r is close to $+1$ or -1 then there is a good straight line fit to the data. If r is close to 0 then there is not a good linear fit to the data.

Evaluation of *r*

For data points $(x_1, y_1), (x_2, y_2), \ldots, (x_n, y_n)$, if you write

$$S_{xx} = \sum (x_i - \bar{x})^2 = \sum x_i^2 - n\bar{x}^2 = \sum x_i^2 - \frac{(\sum x_i)^2}{n}$$

and

$$S_{yy} = \sum (y_i - \bar{y})^2 = \sum y_i^2 - n\bar{y}^2 = \sum y_i^2 - \frac{(\sum y_i)^2}{n}$$

then $(\text{sd})_x = \sqrt{\text{Variance of } x} = \sqrt{\tfrac{1}{n}\sum (x_i - \bar{x})^2} = \sqrt{\tfrac{1}{n} S_{xx}}$

and $(\text{sd})_y = \sqrt{\tfrac{1}{n} S_{yy}}$

- Notice that $S_{xx} = n \times$ Variance of x.
- The equivalence of these formulae for S_{xx} is a by-product of the proof of the equivalence of the two formulae for variance that was presented in the Extension to Chapter 2.
- The formula $S_{xx} = \sum x_i^2 - \frac{(\sum x_i)^2}{n}$ will be most useful in the calculations to be done in this chapter.

The formula $r = \frac{\text{Cov}(x, y)}{(\text{sd})_x(\text{sd})_y}$ can now be rewritten in terms of S_{xy}, S_{xx} and S_{yy}:

$$\Rightarrow \quad r = \frac{\frac{1}{n} S_{xy}}{\sqrt{\frac{1}{n} S_{xx}}\sqrt{\frac{1}{n} S_{yy}}} = \frac{\frac{1}{n} S_{xy}}{\sqrt{\frac{1}{n^2} S_{xx} S_{yy}}} = \frac{\frac{1}{n} S_{xy}}{\frac{1}{n}\sqrt{S_{xx} S_{yy}}}$$

$$\Rightarrow \quad r = \frac{S_{xy}}{\sqrt{S_{xx} S_{yy}}}$$

This is the formula presented in the main text for the correlation coefficient.

Coding does not Affect the Value of the Correlation Coefficient

If data values (x_i, y_i) are transformed to (u_i, v_i) by linear rules

$$u_i = ax_i + b \qquad v_i = cy_i + d \qquad \text{where } a > 0 \text{ and } c > 0$$

then

the correlation coefficient of u and v is the **same** as the correlation coefficient of x and y.

Proof: From your one-variable coding work you know that the coding $u = ax + b$

gives $\bar{u} = a\bar{x} + b$

and $S_{uu} = n \times \text{Variance of } u = n \times a^2 \times \text{Variance of } x = a^2 \times n \times \text{Variance of } x = a^2 \times S_{xx}$

Similarly, the coding $v = cy + d$

gives $\bar{v} = c\bar{y} + d$

and $S_{vv} = c^2 S_{yy}$

Now

$$\begin{aligned} S_{uv} &= \Sigma\,(u - \bar{u})(v - \bar{v}) = \Sigma\,(ax + b - (a\bar{x} + b))(cy + d - (c\bar{y} + d)) \\ &= \Sigma\,(ax - a\bar{x})(cy - c\bar{y}) \\ &= ac\,\Sigma\,(x - \bar{x})(y - \bar{y}) \\ &= acS_{xy} \end{aligned}$$

If r_{uv} denotes the correlation coefficient of the (u, v) data and r_{xy} denotes the correlation coefficient of the (x, y) data then

$$\begin{aligned} r_{uv} &= \frac{S_{uv}}{\sqrt{S_{uu}S_{vv}}} \\ &= \frac{acS_{xy}}{\sqrt{a^2 S_{xx} c^2 S_{yy}}} \\ &= \frac{ac}{\sqrt{a^2c^2}} \frac{S_{xy}}{\sqrt{S_{xx}S_{yy}}} \\ &= \frac{ac}{\sqrt{a^2c^2}} r_{xy} \end{aligned}$$

If a and c are both positive (or both negative) then $\frac{ac}{\sqrt{a^2c^2}} = 1$ so $r_{uv} = r_{xy}$ as required.

EXTENSION 2

Justification of the Formula for the y on x Regression Line

An Observation Based on Intuition

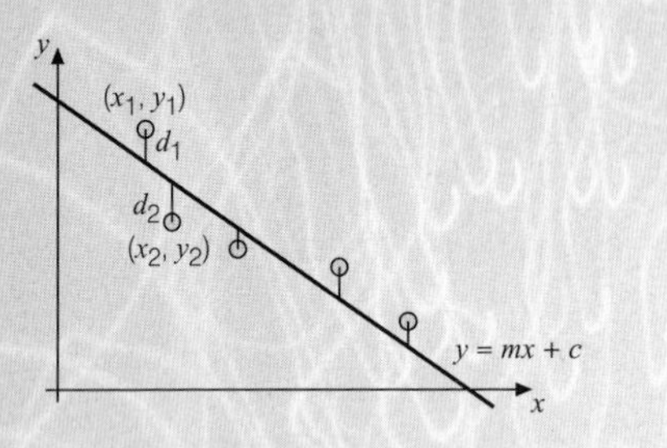

Recall that if the line $y = mx + c$ is used to fit the data points then the vertical deviation of a data point from the line is given by

$$d_i = y_i - (mx_i + c)$$

If $d_i > 0$ then the data point is above the line while if $d_i < 0$ then the data point is below the line.

It makes sense to expect that, for a line that fits the data well, the sum of these deviations should be zero. In other words, it is hoped that the over-estimates and under-estimates cancel each other out.

Algebraically this condition is

$$\begin{aligned} & \Sigma d_i = 0 \\ \Rightarrow \quad & \Sigma (y_i - (mx_i + c)) = 0 \\ \Rightarrow \quad & \Sigma (y_i - mx_i - c) = 0 \\ \Rightarrow \quad & \Sigma y_i - m\Sigma x_i - c\Sigma 1 = 0 \\ \Rightarrow \quad & n\bar{y} - mn\bar{x} - nc = 0 \\ \Rightarrow \quad & \bar{y} - m\bar{x} - c = 0 \quad \text{or} \quad \bar{y} = m\bar{x} + c \end{aligned}$$

If there are n data points then

$$\bar{y} = \frac{\Sigma y}{n} \Rightarrow \Sigma y = n\bar{y}$$

Similarly

$$\Sigma x = n\bar{x}$$

We also have

$$\Sigma 1 = n$$

which means that the point $(\bar{x}, \bar{y})$ must lie on the line.

Any reasonable attempt at a best fit line will pass through the point $(\bar{x}, \bar{y})$.

Finding the Equation of the y on x Line

For data points $(x_1, y_1), (x_2, y_2), \ldots, (x_n, y_n)$ the aim is to find the line that minimises the sum of the squares of the vertical deviations. In other words, you need to find the values of m and c which minimise the value of

$$E = \Sigma d_i^2 = \Sigma (y_i - (mx_i + c))^2$$

You have just seen that any reasonable best fit line will pass through $(\bar{x}, \bar{y})$ so the values m and c must be linked by

$$\begin{aligned} & \bar{y} = m\bar{x} + c \\ \Rightarrow \quad & c = \bar{y} - m\bar{x} \end{aligned}$$

Substituting for c in the expression for E gives

$$E = \Sigma\,(y_i - (mx_i + \bar{y} - m\bar{x}))^2$$
$$= \Sigma\,((y_i - \bar{y}) - m(x_i - \bar{x}))^2$$
$$= \Sigma\,((y_i - \bar{y})^2 - 2m(y_i - \bar{y})(x_i - \bar{x}) + m^2(x_i - \bar{x})^2)$$
$$= \Sigma\,(y_i - \bar{y})^2 - 2m\,\Sigma\,(y_i - \bar{y})(x_i - \bar{x}) + m^2\,\Sigma\,(x_i - \bar{x})^2$$

Expanding the square

If you now recall the notation

$$S_{xx} = \Sigma\,(x_i - \bar{x})^2, \qquad S_{yy} = \Sigma\,(y_i - \bar{y})^2, \qquad S_x = \Sigma\,(x_i - \bar{x})(y_i - \bar{y})$$

then you can write

$$E = S_{yy} - 2S_{xy}m + S_{xx}m^2$$

You need to find the value of m which minimises E: this can be done by differentiation.

$$\frac{\mathrm{d}E}{\mathrm{d}m} = -2S_{xy} + 2mS_{xx}$$

$$\frac{\mathrm{d}^2E}{\mathrm{d}m_2} = 2S_{xx}$$

Refer to C1. You have:

$$E = \alpha - \beta m + \gamma m^2$$

so $\dfrac{\mathrm{d}E}{\mathrm{d}m} = -\beta + 2\gamma m$ and $\dfrac{\mathrm{d}^2E}{\mathrm{d}m^2} = 2\gamma$

For a stationary point $\dfrac{\mathrm{d}E}{\mathrm{d}m} = 0$

$$\Rightarrow \quad -2S_{xy} + 2mS_{xx} = 0$$

$$\Rightarrow \quad m = \frac{S_{xy}}{S_{xx}}$$

Since S_{xx} is a sum of squares, it must be positive. $\dfrac{\mathrm{d}^2E}{\mathrm{d}m^2}$ is therefore positive which tells you that E does indeed have a minimum value when $m = \dfrac{S_{xy}}{S_{xx}}$.

You now know that the best fit line that minimises the sum of the squares of the vertical distances:

- passes through the point $(\bar{x}, \bar{y})$
- has gradient $\dfrac{S_{xy}}{S_{xx}}$

so it can be said that the best fit line must have equation $y - \bar{y} = \dfrac{S_{xy}}{S_{xx}}(x - \bar{x})$

This is the equation of the y on x line presented earlier in the chapter.

Remember that the line through (a, b) with gradient m has equation $y - b = m(x - a)$

Having studied this chapter you should know how to

- illustrate bivariate data using a scatter diagram
- calculate the product moment correlation coefficient of n data points using the formula

$$r = \frac{S_{xy}}{\sqrt{S_{xx}S_{yy}}}$$

where

$$S_{xx} = \sum(x-\bar{x})^2 = \sum x^2 - \frac{(\sum x)^2}{n}, \quad S_{yy} = \sum(y-\bar{y})^2 = \sum y^2 - \frac{(\sum y)^2}{n}$$

$$S_{xy} = \sum(x-\bar{x})(y-\bar{y}) = \sum xy - \frac{(\sum x)(\sum y)}{n}$$

- interpret calculated values of r
  [In particular, it should be known that $r = 1$ means the data points lie on a line of positive gradient; r being close to 1 means that the data points are close to a line of positive gradient; r being close to -1 means that the data points are close to a line of negative gradient; $r = -1$ means the data points lie on a line of negative gradient; r being close to 0 means that there isn't a good straight line approximation to the data points.
  It should also be known that r has no units and that the value of r is unaffected by linear changes of variable.]
- calculate and use the **y on x least squares regression line** which minimises the sum of the squares of the vertical distances between the data points and the line and has equation $y - \bar{y} = \frac{S_{xy}}{S_{xx}}(x - \bar{x})$
- calculate and use the **x on y least squares regression line** which minimises the sum of the squares of the horizontal distances between the data points and the line and has equation $x - \bar{x} = \frac{S_{xy}}{S_{yy}}(y - \bar{y})$
- obtain predictions from the appropriate regression line
  [If x and y are both random variables then predictions of y values should be obtained from the y on x line and predictions of x should be obtained from the x on y line.
  If the x data is controlled then all predictions should be based on the y on x line.
  If the y data is controlled then all predictions should be based on the x on y line.]
- evaluate the reliability of predictions made from regression lines
  [A prediction will generally be reliable if interpolation is being used and the value of r is close to 1 or -1.]

REVISION EXERCISE

1 A set of bivariate data (x, y) was collected for an experiment. You are given that $n = 10$, $\sum x = 78.9$, $\sum y = 64.0$, $\sum x^2 = 743.45$, $\sum y^2 = 476.52$, $\sum xy = 575.81$.
You are also given that the regression line of x on y has equation $x = 1.11 + 1.06y$, where the coefficients are given correct to 3 significant figures.

i) Calculate the equation of the regression line of y on x, giving your answer in the form $y = a + bx$.

ii) State the co-ordinates of the point of intersection of the two lines. (You do not need to solve the simultaneous equations.)

iii) In the collection of the data neither variable was controlled. Use the appropriate line to estimate the value of x when $y = 3.4$. You may assume that $y = 3.4$ is within the range of the data collected.

2 A football was rolled in a straight line along the floor of a sports hall. The distance travelled, x metres, was recorded at a time, t seconds, after the ball was released, for values of t from $t = 1.0$ to $t = 4.5$. The results are given in the table below:

t	1.0	1.5	2.0	2.5	3.0	3.5	4.0	4.5
x	8.0	13.2	15.8	18.4	21.0	22.2	24.0	25.1

$[n = 8, \sum t = 22.0, \sum x = 147.7, \sum t^2 = 71.00, \sum x^2 = 2966.29, \sum tx = 455.05]$

i) On graph paper plot a scatter diagram of the data.

ii) Calculate the product moment correlation coefficient for the data.

iii) State with a reason whether a linear model would be appropriate for these data, referring to either your calculations or the scatter diagram.

3 In a science experiment a student drops a ball on to the ground from a height of x metres and measures the height, y metres, to which it bounces. The experiment is repeated to give 10 pairs of results. The student chooses values of x that increase in steps of exactly 0.25 metres. The results of the experiment are shown in the table below:

x	0.25	0.50	0.75	1.00	1.25	1.50	1.75	2.00	2.25	2.50
y	0.11	0.17	0.29	0.38	0.46	0.54	0.63	0.71	0.84	0.90

i) Calculate the product moment correlation coefficient between x and y.

ii) By calculating the equation of the appropriate regression line, estimate the height to which the ball will bounce if it is dropped from a height of 2.4 metres.

iii) Comment on the reliability of:

a) the estimate found in part (ii);

b) the estimate obtained from the regression line of the height to which the ball will bounce if it is dropped from a height of 10.2 metres.

(OCR Jun 2001 S1)

4 The table below shows the quantity, x units, of electricity produced by nuclear power and the quantity, y units, of electricity produced by hydro-electric power, for the UK over a period of 12 months.

Month	J	F	M	A	M	J	J	A	S	O	N	D
x	1.80	1.60	2.00	1.60	1.70	2.00	1.51	1.65	1.97	1.55	1.68	2.11
y	0.05	0.04	0.07	0.05	0.04	0.04	0.02	0.02	0.02	0.02	0.04	0.06

$[n = 12, \sum x = 21.17, \sum y = 0.47, \sum x^2 = 37.8105, \sum y^2 = 0.0215, \sum xy = 0.8494]$

A researcher believes there may be a link between the variables x and y.

i) Calculate the product moment correlation coefficient for the data.

ii) Calculate the equation of the regression line of y on x, giving your answer in the form $y = a + bx$.

iii) Use your regression line to estimate the amount of electricity produced by hydro-electric power in a month in which the quantity of electricity produced by nuclear power is 2.52.

iv) Comment on the reliability of the estimate found in part (iii)

(OCR Jan 2001 S1)

5

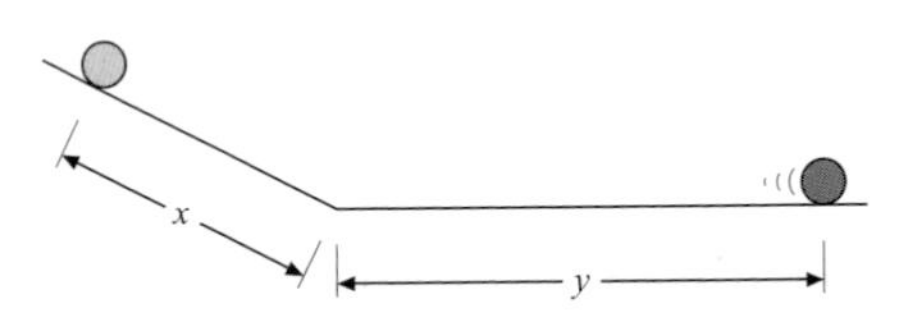

A student conducted an experiment in which she rolled a ball from a point on an inclined plane. The ball rolled down the inclined plane and then continued to roll along a horizontal plane (see diagram). The student measured the distance, x cm, up the inclined plane to the ball's initial position. She also measured the distance, y cm, travelled by the ball along the horizontal plane before it stopped. She repeated the experiment to give 10 pairs of readings altogether. The values of x were increased at regular intervals of 5 cm starting at 40 cm. For each repetition of the experiment she tried to roll the ball down the plane with the same initial speed. The results are given in the table below:

x	40	45	50	55	60	65	70	75	80	85
y	58	69	70	73	88	91	98	108	109	125

$[n = 10, \sum x = 625, \sum y = 889, \sum x^2 = 41\ 125, \sum y^2 = 83\ 153, \sum xy = 58\ 440]$

i) Calculate the value of the product moment correlation coefficient for the data.

ii) The student's teacher suggested that, instead of using the original x and y values, she should transform the data using the equations $u = \frac{x-40}{5}$ and $v = y - 50$, and then calculate the product moment correlation coefficient for the transformed data. Explain what relationship, if any, the product moment correlation coefficient for the transformed data would have with the product moment correlation coefficient calculated in part (i).

The student wishes to estimate the value of x for a ball which rolls a horizontal distance of $y = 100$ cm.

ii) Calculate the equation of an appropriate regression line and use it to estimate the value of x when $y = 100$.

iv) Give a reason for your choice of regression line.

(OCR Jan 2002 S1)

6 The three scatter diagrams show how three variables, u, v and w, vary with respect to a fourth variable x:

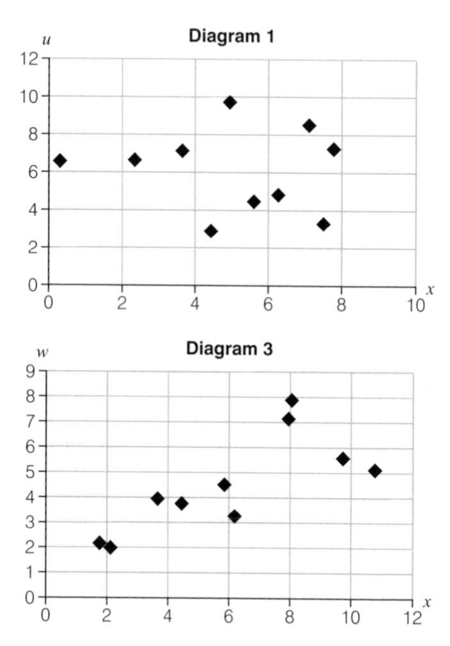

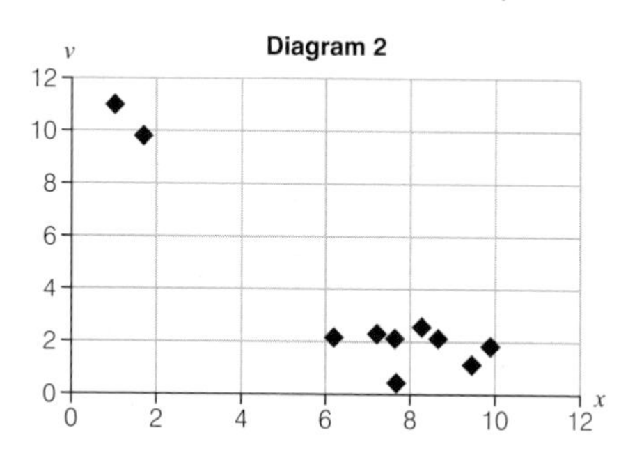

The product moment correlation coefficients for each of the three sets of data have been calculated as

0.760, –0.101, –0.936

State, with a reason in each case, which correlation coefficient corresponds to each of the diagrams 1, 2 and 3.

7 The relationship between variables x and y has been investigated by collecting three pairs of values of (x, y). The values were (3, 4), (5, 10) and (7, 25).

a) Calculate the product moment correlation coefficient for this data.

The data points are shown in the scatter diagram, together with a straight line fitted by eye. The equation of this line is $y = 4x - 6$.

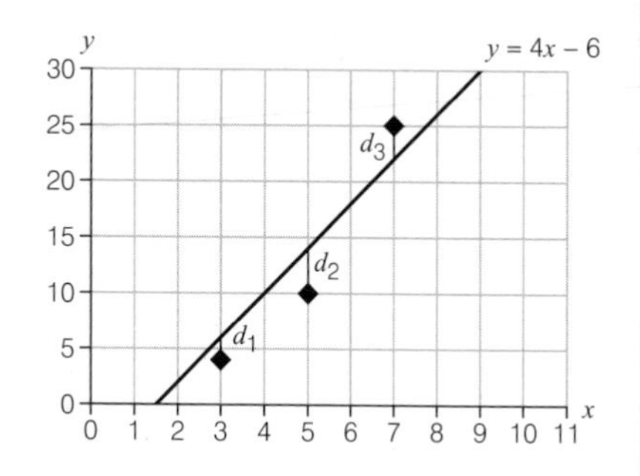

b) Calculate the sum of the squares of the vertical deviations of the three data points from the straight line fitted by eye. (That is, calculate the value of $d_1^2 + d_2^2 + d_3^2$.)

c) Calculate the equation of the least squares regression line of y on x for the data points.

d) Verify that the sum of the squares of the vertical deviations for this line is smaller than the sum of the squares of the vertical deviations for the line $y = 4x - 6$.

8 A set of bivariate data (x, y) was collected in an experiment in which x is the yield, in tonnes, of a particular crop in a given month and y is the number of centimetres of rainfall in that month. You are told that

$$n = 6, \sum x = 46.7, \sum y = 42.5, \sum x^2 = 400.65, \sum y^2 = 323.67, \sum xy = 346.37$$

a) Calculate the product moment correlation coefficient for the data.

b) Calculate the equation of the regression line of x on y, giving your answer in the form $x = a + by$.

c) Use the equation found in part (b) to estimate the yield of the crop in a month in which the rainfall was 20.4 cm.

d) State with a reason whether you think the estimate found in part (b) will be reliable.

(OCR Nov 2002 S1)

9 Obtain the product movement correlation coefficient for the bivariate data summarised by

$$n = 8, \sum x = 22.4, \sum y = 52.0, \sum x^2 = 69.44, \sum y^2 = 452.24, \sum xy = 138.88$$

It is known that, in fact, there is a very strong relationship between x and y. State with a reason whether this relationship is linear or non-linear.

(OCR Jun 2000 S1)

10 The following table gives the index numbers of the output of export industries, x, and of agriculture, y, during various years of the eighteenth century:

Year	Export x	Agriculture y
1700	100	100
1710	108	104
1720	125	105
1730	142	103
1740	148	104
1750	176	111
1760	222	115
1770	256	117
1780	246	126
1790	383	135
1800	544	143

$[n = 11, \sum x = 2450, \sum y = 1263, \sum x^2 = 728\,294, \sum y^2 = 147\,051, \sum xy = 299\,886]$

i) Obtain the equation of the regression line of y on x, giving your answer in the form $y = a + bx$.

ii) It is desired to estimate the value of x corresponding to $y = 140$. Discuss the suitability of the line of regression of y on x for this estimate. You should refer in your answer to the value $y = 140$, the choice of regression line and the value of the product moment correlation coefficient between x and y, which is 0.96 correct to 2 significant figures.

(OCR Jun 2000 S1)

11 The resistive force, y Newtons, acting on a motor-cyclist depends on the motor-cyclist's speed, x km per hour. The following table shows the values of x corresponding to various values of x:

x	10	20	40	50	60	70	80	100
y	70	85	115	130	168	174	268	332

i) Plot a scatter diagram to illustrate the data.

ii) State the value of the product moment correlation coefficient for the points corresponding to $x = 10$, 20, 40 and 50.

iii) Calculate the equation of an appropriate line of regression for the points corresponding to $x = 50$, 60, 70, 80 and 100, simplifying your answer as far as possible.

[For these points, $n = 5$, $\sum x = 360$, $\sum y = 1072$, $\sum x^2 = 27\ 400$, $\sum y^2 = 257\ 448$, $\sum xy = 83\ 400$.]

iv) Estimate the values of y corresponding to $x = 30$ and $x = 90$. Compare the reliability of the two estimated values.

(OCR Mar 2000 S1)

12 Twelve candidates took two examination papers, one in English and one in Mathematics. Their marks are shown in the table:

Candidate	A	B	C	D	E	F	G	H	I	J	K	L
English mark (x)	22	26	28	34	42	51	54	59	60	70	74	81
Mathematics mark (y)	3	5	10	69	47	50	76	61	35	97	98	100

[$n = 12$, $\sum x = 601$, $\sum y = 651$, $\sum x^2 = 34\ 399$, $\sum y^2 = 49\ 339$, $\sum xy = 39\ 291$]

Candidate W scored 35 on the Mathematics paper but was absent from the English paper. It is desired to use a line of regression to find an estimated mark for W in English.

i) Calculate the equation of an appropriate line of regression and use it to obtain an estimate of W's English mark.

ii) Calculate the product moment correlation coefficient for the 12 candidates A to L.

iii) The scatter diagram for the 12 candidates is shown. State what feature of the diagram illustrates:

a) your answer to part (ii);

b) the fact that the value of the product moment correlation coefficient for the six candidates D, E, F, G, H and I is –0.29, correct to 2 significant figures.

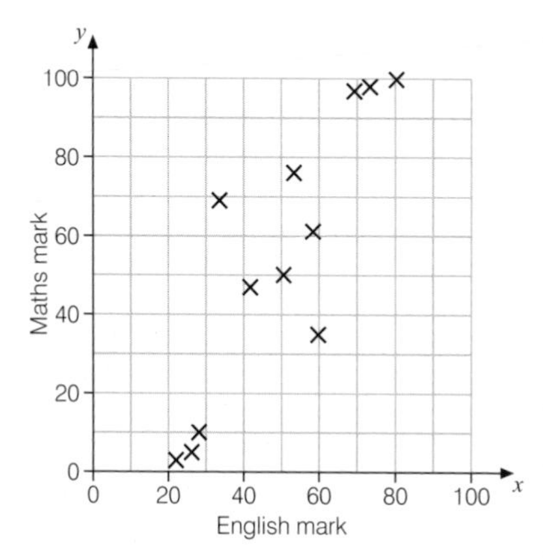

(OCR Jun 1999 S1)

11 Rank Correlation

The purpose of this chapter is to enable you to

- calculate the product moment correlation coefficient for data consisting of ranks using both the usual formulae for the correlation coefficient and the alternative formulae for ranked data
- use Spearman's rank correlation coefficient as a quickly calculated correlation coefficient and realise it often gives a good approximation to the product moment correlation coefficient

The Product Moment Correlation Coefficient for Ranks

A frequently occurring example of bivariate data is the situation where the data consists of the positions, or ranks, assigned by two judges to the participants in a competitive event.

Suppose, for example, a "beautiful baby" competition has eight entrants and the ranking decisions of the two judges are shown in the table below (1 denotes the most beautiful baby, 2 the second most beautiful, etc.):

	Baby A	Baby B	Baby C	Baby D	Baby E	Baby F	Baby G	Baby H
Judge 1	3	4	7	2	8	5	1	6
Judge 2	6	2	8	1	7	5	3	4

This data could be illustrated in a scatter diagram and the correlation coefficient calculated:

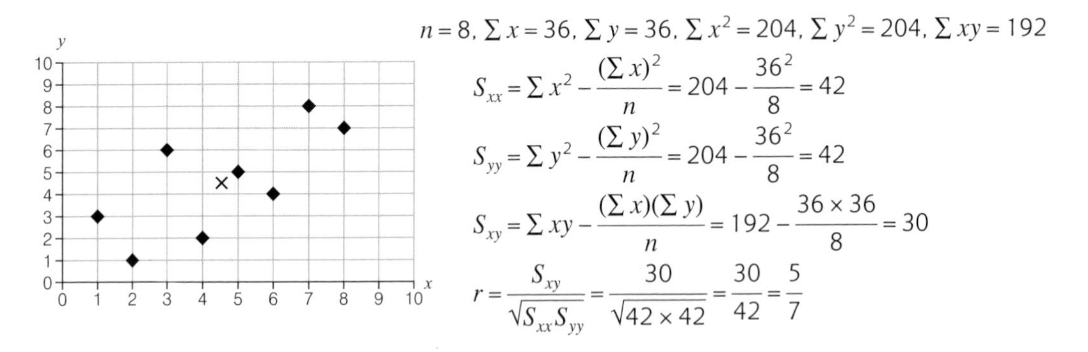

$n = 8, \sum x = 36, \sum y = 36, \sum x^2 = 204, \sum y^2 = 204, \sum xy = 192$

$$S_{xx} = \sum x^2 - \frac{(\sum x)^2}{n} = 204 - \frac{36^2}{8} = 42$$

$$S_{yy} = \sum y^2 - \frac{(\sum y)^2}{n} = 204 - \frac{36^2}{8} = 42$$

$$S_{xy} = \sum xy - \frac{(\sum x)(\sum y)}{n} = 192 - \frac{36 \times 36}{8} = 30$$

$$r = \frac{S_{xy}}{\sqrt{S_{xx}S_{yy}}} = \frac{30}{\sqrt{42 \times 42}} = \frac{30}{42} = \frac{5}{7}$$

You can see from the scatter diagram and the correlation coefficient that there is reasonable agreement between the ranks the two judges have given the babies.

If there was perfect agreement between the two judges, the ranks given by judge 2 would be exactly the same as the ranks given by judge 1.

The data points would all lie on the line $y = x$ so the correlation coefficient would be 1.

On the other hand if there was complete disagreement between the two judges then

the baby placed 1st by judge 1 would be placed 8th by judge 2;

the baby placed 2nd by judge 1 would be placed 7th by judge 2;

the baby placed 3rd by judge 1 would be placed 6th by judge 2;

the baby placed 4th by judge 1 would be placed 5th by judge 2;

etc.

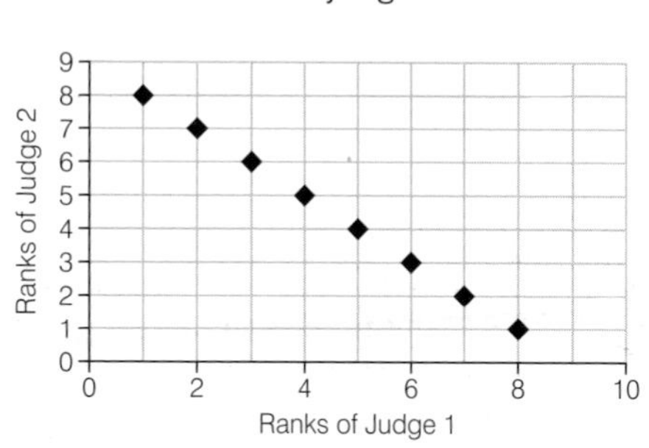

The data points (1, 8), (2, 7), (3, 6), (4, 5), etc. all lie on the line $y = 9 - x$. This is a line of negative gradient so the correlation coefficient for judges in complete disagreement would be -1.

For data pairs that are ranks

- a correlation coefficient of 1 signifies complete agreement between the two judges;
- a correlation coefficient of -1 signifies complete disagreement between the two judges.

EXERCISE 1

1 Three judges were asked to judge a photograph competition. In the final round of the competition there were just six photos. The rank orders for the photos given by the three judges are shown in the table below; 1 signifies the best photo and 6 the worst.

	Photo A	Photo B	Photo C	Photo D	Photo E	Photo F
Judge A	1	2	3	4	5	6
Judge B	2	3	1	4	6	5
Judge C	5	4	3	6	1	2

a) Draw a scatter diagram to show the relationship between the ranks of judge A and judge B. Calculate the correlation coefficient for these ranks.
b) Draw a scatter diagram to show the relationship between the ranks of judge B and judge C. Calculate the correlation coefficient for these ranks.
c) Draw a scatter diagram to show the relationship between the ranks of judge C and judge A. Calculate the correlation coefficient for these ranks.

What conclusions can be drawn about the ranks given by the three judges?

2 At a flower show two judges were asked to rank eight displays. The results are shown below with rank 1 being given to the display the judge likes the most.

	Display A	Display B	Display C	Display D	Display E	Display F	Display G	Display H
Judge 1	1	2	3	4	5	6	7	8
Judge 2	8	2	5	4	3	6	7	1

Draw a scatter diagram to show the relationship between the ranks given by the two judges. Calculate the correlation coefficient for these ranks. What conclusions can you draw about the ranks given by the two judges?

Alternative Formula for the Evaluation of the Correlation Coefficient of Ranks

In the Appendix to this chapter, a much simpler way of calculating the correlation coefficient of data consisting of ranks is derived.

If the n data values $x_1, x_2, \ldots, x_n$ are simply the values 1, 2, 3, ..., n taken in any order and the data values $y_1, y_2, \ldots, y_n$ are simply the values 1, 2, 3, ..., n taken in any order then the correlation coefficient for data pairs $(x_1, y_1), (x_2, y_2), \ldots, (x_n, y_n)$ can be calculated using the formula

$$r = 1 - \frac{6\sum d_i^2}{n(n^2-1)} \quad \text{where} \quad d_i = x_i - y_i$$

You can check this formula and see how quick it is to use by referring back to the set of data considered at the beginning of the chapter:

	Baby A	Baby B	Baby C	Baby D	Baby E	Baby F	Baby G	Baby H	
Judge 1	3	4	7	2	8	5	1	6	
Judge 2	6	2	8	1	7	5	3	4	
d_i^2	9	4	1	1	1	0	4	4	$\sum d^2 = 24$

$(3-6)^2 = 9$

$(4-2)^2 = 4$

You have $n = 8$ so $r = 1 - \frac{6\sum d_i^2}{n(n^2-1)} = 1 - \frac{6 \times 24}{8 \times 63} = 1 - \frac{2}{7} = \frac{5}{7}$

EXAMPLE 1

A class of 12 pupils wish to compare their performances in a History exam and a Physics exam. The table below shows the class positions of each pupil in the two subjects:

Pupil	A	B	C	D	E	F	G	H	I	J	K	L
History	1	2	3	4	5	6	7	8	9	10	11	12
Physics	3	11	8	9	7	2	5	6	4	12	1	10

Obtain and comment on the correlation coefficient of this data.

SOLUTION

Pupil	A	B	C	D	E	F	G	H	I	J	K	L	
History	1	2	3	4	5	6	7	8	9	10	11	12	
Physics	3	11	8	9	7	2	5	6	4	12	1	10	
d^2	4	81	25	25	4	16	4	4	25	4	100	4	$\Sigma d^2 = 296$

You have $n = 12$ so $r = 1 - \dfrac{6 \Sigma d_i^2}{n(n^2 - 1)} = 1 - \dfrac{6 \times 296}{12 \times 143} = 1 - 1.03496... = -0.0350$ (3 s.f.)

There is no clear linear correlation between the pupil's positions in History and their positions in Physics. This is confirmed by drawing a scatter diagram of the data:

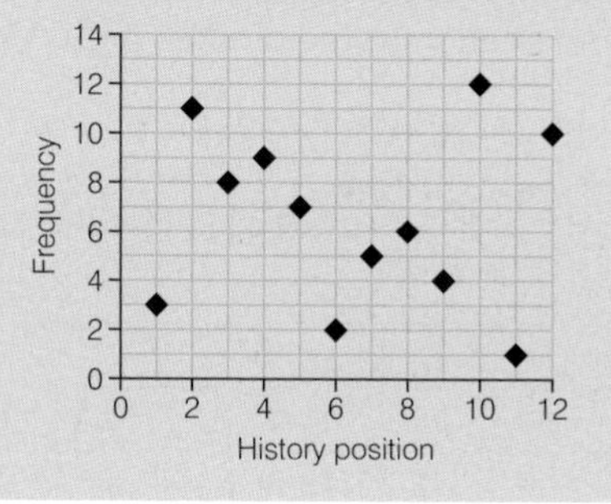

EXERCISE 2

1 Two teenagers decided to each place five musical groups into order with 1 representing their favourite group and 5 representing their least favourite group.
The results were:

	Group P	Group Q	Group R	Group S	Group T
Anne's ranking	1	4	2	5	3
Belinda's ranking	2	5	1	4	3

Use the $r = 1 - \dfrac{6 \Sigma d_i^2}{n(n^2 - 1)}$ formula to calculate the correlation coefficient of this data.

2 A football league has 8 teams playing in it. The table below shows the final positions of the teams at the end of the 2002/3 season and at the end of the 2003/4 season.

	Team A	Team B	Team C	Team D	Team E	Team F	Team G	Team H
2002/3 position	1	5	2	3	7	8	4	6
2003/4 position	1	8	4	2	3	5	6	7

Draw a scatter diagram to illustrate the data and calculate the correlation coefficient of this data. What conclusions can be drawn from your diagram and calculation?

Spearman's Coefficient of Rank Correlation

This calculation for ranked data is so easy compared with the usual calculation of the product moment correlation coefficient for other sets of data that it is sometimes used **for any set of data:**

- the x values are first ranked (with the highest being ranked number 1, the second highest number 2, ..., the lowest being ranked n);
- the y values are ranked in the same way;
- **the correlation coefficient of the resulting ranks is then calculated**.

The result is called **Spearman's coefficient of rank correlation,** r_s, which will often, but not always, give a reasonable approximation to the product moment correlation coefficient.

EXAMPLE 2

Ten students were selected at random from a class and their percentage marks in Chemistry and Art exams were recorded:

Pupil	1	2	3	4	5	6	7	8	9	10
Chemistry %	80	48	62	53	45	72	60	41	52	57
Art %	53	61	56	72	48	51	45	58	50	46

Calculate the value of Spearman's coefficient of rank correlation for the data.

SOLUTION

First rank the data and then calculate the values of d^2:

Pupil	1	2	3	4	5	6	7	8	9	10	
Chem %	80	48	62	53	45	72	60	41	52	57	
Art %	53	61	56	72	48	51	45	58	50	46	
Chem rank	**1**	**8**	**3**	**6**	**9**	**2**	**4**	**10**	**7**	**5**	
Art rank	**5**	**2**	**4**	**1**	**8**	**6**	**10**	**3**	**7**	**9**	
d^2	16	36	1	25	1	16	36	49	0	16	$\Sigma d^2 = 196$

You have $n = 10$ so $r_s = 1 - \dfrac{6\Sigma d_i^2}{n(n^2-1)} = 1 - \dfrac{6 \times 196}{10 \times 99} = -0.188$ (3 d.p.)

If the product moment correlation coefficient is calculated for the **original percentage marks** in the two subjects, a value of $r = -0.217$ (3 d.p.) is obtained.

You can see that, in this case, r_s is a reasonable approximation to r.

EXAMPLE 3

As part of an interview process, the five candidates were asked to complete two tests: an intelligence test and an agility test. The scores of the candidates are shown in the table below.

	Candidate P	Candidate Q	Candidate R	Candidate S	Candidate T
Intelligence test score (x)	32	35	41	38	14
Agility test score (y)	29	27	26	25	11

a) Given that

$\sum x = 160, \sum y = 118, \sum x^2 = 5570, \sum y^2 = 2992, \sum xy = 4043$

calculate the product moment correlation coefficient of the data.

b) Calculate Spearman's rank correlation coefficient for the data.

c) By referring to the scatter diagram for the original data and for the ranks, comment on your results.

SOLUTION

a) $$S_{xx} = \sum x^2 - \frac{(\sum x)^2}{n} = 5570 - \frac{160^2}{5} = 450$$

$$S_{yy} = \sum y^2 - \frac{(\sum y)^2}{n} = 2992 - \frac{118^2}{5} = 207.2$$

$$S_{xy} = \sum xy - \frac{(\sum x)(\sum y)}{n} = 4043 - \frac{160 \times 118}{5} = 267$$

$$r = \frac{S_{xy}}{\sqrt{S_{xx}S_{yy}}} = \frac{267}{\sqrt{450 \times 207.2}} = 0.874 \quad \text{(3 d.p.)}$$

b)

	P	Q	R	S	T	
Intelligence test score (x)	32	35	41	38	14	
Agility test score (y)	29	27	26	25	11	
Intelligence ranks	**4**	**3**	**1**	**2**	**5**	
Agility ranks	**1**	**2**	**3**	**4**	**5**	
d^2	9	1	4	4	0	$\sum d^2 = 18$

To calculate Spearman's coefficient remember that you must first obtain the ranks of the data.

You have $n = 5$ so $r_s = 1 - \frac{6\sum d_i^2}{n(n^2-1)} = 1 - \frac{6 \times 18}{5 \times 24} = 0.1$.

c) In this case the values of the product moment coefficient of correlation and Spearman's coefficient of correlation are markedly different.

EXAMPLE 3 (continued)

The scatter diagram for the original data shows that four sets of results are very similar to each other whilst one result is very different. The "cloud" of four results and one outlier is enough to produce a positive product moment correlation coefficient.

On the other hand, the scatter diagram for the **ranks** of the two sets of scores makes it clear that there is no clear pattern in the rank performances of the candidates.

This example should again remind you of the wisdom of drawing scatter diagrams as well as calculating correlation coefficients!

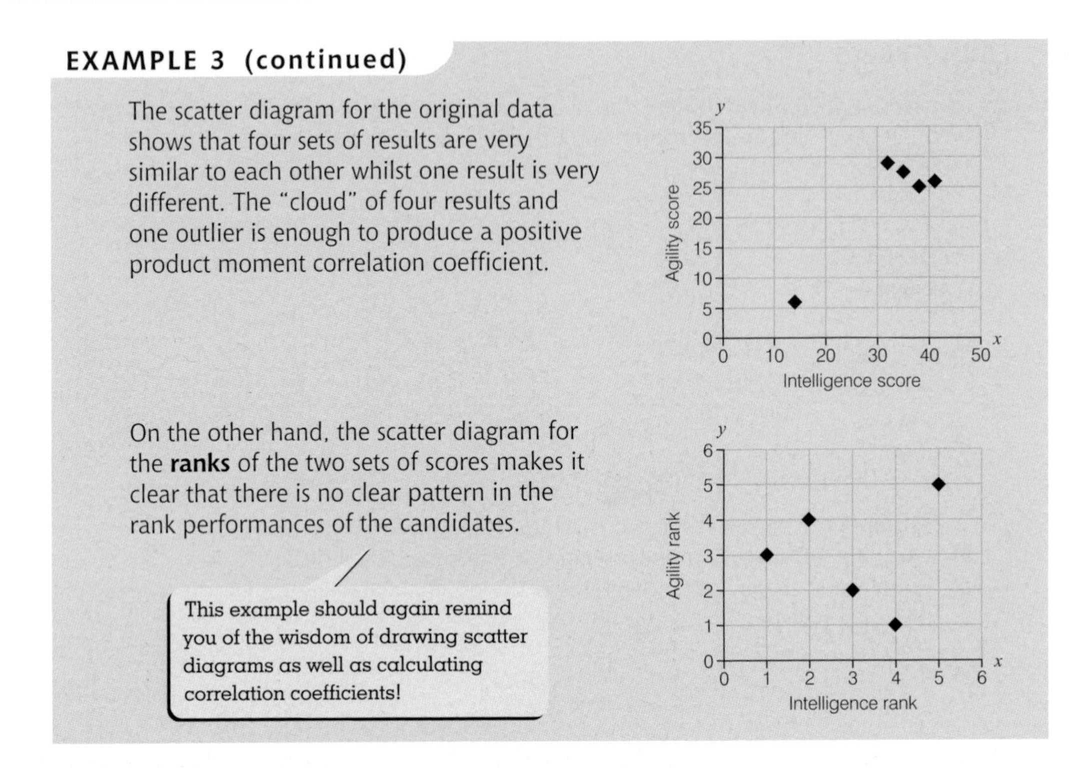

Interpreting Values of 1 and −1 for r_s

If the value of r_s is 1 then the ranks must be in perfect agreement: the data point (x_i, y_i) with the lowest x value must also have the lowest y value; the data point (x_i, y_i) with the second lowest x value must also have the second lowest y value; the data point (x_i, y_i) with the third lowest x value must also have the third lowest y value, etc. This means that the scatter graph of the data points must be steadily increasing. A possible scatter graph for data with $r_s = 1$ is shown in the diagram. Note that in this case $r \neq 1$ since the points do not lie on a line.

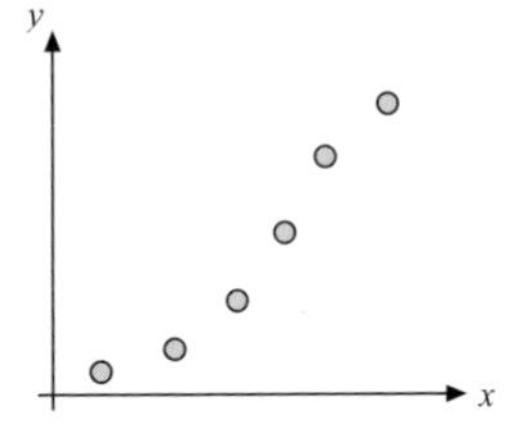

Similarly, **if $r_s = -1$ then the ranks must be in complete disagreement**: the data point (x_i, y_i) with the lowest x value must have the highest y value; the data point (x_i, y_i) with the second lowest x value must have the second highest y value, etc. This means that the scatter diagram of the data points must be steadily decreasing.

A possible scatter diagram for data points with $r_s = -1$ is shown in the diagram.

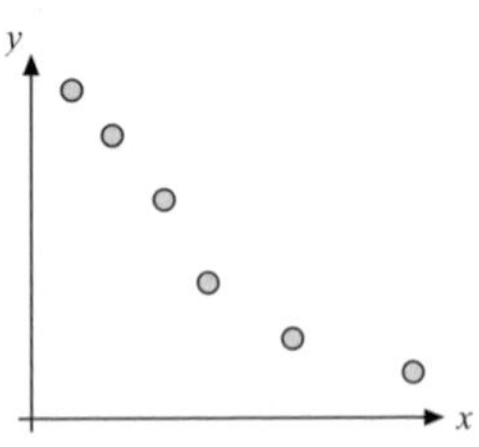

EXERCISE 3

1 Two friends watched the same TV programmes during a weekend and gave each programme a score out of 20. Their results were:

Programme	A	B	C	D	E	F	G	H
Anne's score	17	11	13	18	6	5	19	12
Betty's score	14	10	13	17	9	11	16	15

a) Calculate the product moment coefficient of correlation for the two sets of scores.
b) Calculate Spearman's coefficient of rank correlation for the two sets of data.

2 Ten pupils had two pieces of work to do one evening. They had to draw a map for Geography and write a letter for English. The table below shows the times, in minutes, that they spent on each of these tasks.

Pupil	A	B	C	D	E	F	G	H	I	J
Time on map (mins)	32	47	35	21	28	42	49	26	36	15
Time on letter (mins)	35	44	38	24	26	45	47	22	30	19

Calculate Spearman's coefficient of rank correlation for the two sets of times.

3 The exam marks for a group of eight students in two exam papers are shown in the table below:

Pupil	A	B	C	D	E	F	G	H
Paper 1	83	58	56	53	51	48	45	44
Paper 2	89	35	36	38	40	41	44	48

a) Draw a scatter diagram for the marks and calculate the product moment coefficient of linear correlation for the data.
b) Calculate Spearman's coefficient of rank correlation for the data. Draw a scatter diagram of the ranks in each paper.

Comment on the statement that "r_s is often used as a quick means of estimating the product moment correlation coefficient".

4 It is believed that a patient who absorbs a drug well on one occasion will do so on another occasion. Ten patients gave the following results for the percentage absorbed on two days:

Patient	A	B	C	D	E	F	G	H	I	J
Day 1	35.5	16.6	13.6	42.5	39.0	29.5	28.5	36.0	19.7	42.0
Day 2	27.6	15.1	12.9	30.5	23.1	14.5	35.5	27.5	16.1	18.9

Calculate a rank correlation coefficient for the data and use it to decide whether the belief appears justified.

5 Records of annual wheat crop in a farming area and total rainfall during June, July, August for six years were as follows:

Year	1	2	3	4	5	6
Yield (megatons)	8.3	9.2	7.3	8	5.2	8.2
Summer rainfall (cm)	22	25	18	19	24	16

Calculate:

a) the product moment correlation coefficient

b) a rank correlation coefficient

between the rainfall and the crop.

State briefly the advantages of each of these measures of correlation.

6 Draw, where possible, a sketch of a scatter diagram for a set of data where:

a) the product moment correlation coefficient is 1;

b) Spearman's coefficient of rank correlation is 1 but the product moment correlation coefficient is not 1;

c) Spearman's coefficient of rank correlation is -1 and the product moment correlation coefficient is -1;

d) Spearman's coefficient of rank correlation is not 1 but the product moment correlation coefficient is 1;

e) the product moment correlation coefficient is approximately zero.

EXTENSION

Development of the Alternative Formula for the Correlation Coefficient for Data Consisting of Ranks

In question 1 of Exercise 1 you calculated three different correlation coefficients for the ranks given by two judges to six competitors. You probably noticed that many of the details of each calculation turned out to be the same.

Make sure you have worked through Q1 of Exercise 1 before reading through this section.

In each case

$$n = 6,\ \sum x = 21,\ \sum y = 21,\ \sum x^2 = 91,\ \sum y^2 = 91$$

The only thing that varied from example to example was $\sum xy$.

The value of n was 6 in each case because you were considering the ranks given to six competitors.

The value of $\sum x$ was 21 in each case since the ranks given by the first judge to be considered must be the numbers 1, 2, 3, 4, 5, 6 in some order and these numbers add to 21. The same argument tells you that $\sum y = 21$ for each example.

Similarly $\sum x^2 = 91 = \sum y^2$ since both are simply the sum of $1^2, 2^2, 3^2, 4^2, 5^2$ and 6^2 in some order.

Similar results hold for other numbers of competitors. You can tabulate these results and search for a general formula:

Number of competitors	$\sum x = \sum y =$	$\sum x^2 = \sum y^2 =$
3	$1+2+3=\mathbf{6}$	$1^2+2^2+3^2=\mathbf{14}$
4	$1+2+3+4=\mathbf{10}$	$1^2+2^2+3^2+4^2=\mathbf{30}$
5	$1+2+3+4+5=\mathbf{15}$	$1^2+2^2+3^2+4^2+5^2=\mathbf{55}$
6	21	91
7	28	140
8	36	204
9	45	285
10	55	385
n	$\frac{1}{2}n(n+1)$	

$1+2+3+\cdots+n$
is an arithmetic progression with first term 1, last term n and it has n terms.

Using the result $\frac{1}{2}n$(first term + last term) for the sum of an arithmetic progression you have $1+2+3+\cdots+n=\frac{1}{2}n(1+n)$

To find a rule for $\sum x^2$ look at the table again and consider the multiplying factor to move from $\sum x$ to $\sum x^2$:

Number of competitors	$\sum x = \sum y =$		$\sum x^2 = \sum y^2 =$
3	**6**	$\xrightarrow{\times\frac{7}{3}}$	**14**
4	**10**	$\xrightarrow{\times\frac{9}{3}}$	**30**
5	**15**	$\xrightarrow{\times\frac{11}{3}}$	**55**
6	21	$\xrightarrow{\times\frac{13}{3}}$	91
7	28	$\xrightarrow{\times\frac{15}{3}}$	140
8	36	$\xrightarrow{\times\frac{17}{3}}$	204
9	45	$\xrightarrow{\times\frac{19}{3}}$	285
10	55	$\xrightarrow{\times\frac{21}{3}}$	385
n	$\frac{1}{2}n(n+1)$		

The multiplying factor seems to be $\frac{2n+1}{3}$ so you can make the conjecture that

$$\sum x^2 = 1^2+2^2+\cdots+n^2 = \frac{1}{2}n(n+1)\times\frac{2n+1}{3} = \frac{1}{6}n(n+1)(2n+1)$$

If you study module FP1 you will probably meet a formal proof of this result.

Now consider

$$S_{xx} = \Sigma x^2 - \frac{(\Sigma x)^2}{n}$$
$$= \tfrac{1}{6}n(n+1)(2n+1) - \frac{(\tfrac{1}{2}n(n+1))^2}{n}$$
$$= \tfrac{1}{6}n(n+1)(2n+1) - \frac{\tfrac{1}{4}n^2(n+1)^2}{n}$$
$$= \tfrac{1}{6}n(n+1)(2n+1) - \tfrac{1}{4}n(n+1)(2n+1)$$
$$= \tfrac{1}{12}n(n+1)(2(2n+1) - 3(n+1))$$

Common factor of $\tfrac{1}{12}n(n+1)$

$$= \tfrac{1}{12}n(n+1)(4n+2-3n-3)$$
$$= \tfrac{1}{12}n(n+1)(n-1)$$
$$= \tfrac{1}{12}n(n^2-1)$$

Similarly $S_{yy} = \tfrac{1}{12}n(n^2-1)$

You can obtain a useful formulation for S_{xy} by first noting that

$$x_i y_i = \tfrac{1}{2}(x_i^2 + y_i^2 - (x_i - y_i)^2) = \tfrac{1}{2}(x_i^2 + y_i^2 - d_i^2)$$

where

$d_i = x_i - y_i =$ difference between ranks given by two judges to competitor i.

So

$$S_{xy} = \Sigma x_i y_i - \frac{(\Sigma x_i)(\Sigma y_i)}{n} = \Sigma \frac{1}{2}(x_i^2 + y_i^2 - d_i^2) - \frac{(\Sigma x_i)(\Sigma y_i)}{n}$$
$$= \frac{1}{2}\Sigma x_i^2 + \frac{1}{2}\Sigma y_i^2 - \frac{1}{2}\Sigma d_i^2 - \frac{(\Sigma x_i)(\Sigma y_i)}{n}$$

For rank data
$\Sigma x = \Sigma y$
and
$\Sigma x^2 = \Sigma y^2$

$$= \Sigma x_i^2 - \frac{1}{2}\Sigma d_i^2 - \frac{(\Sigma x_i)(\Sigma x_i)}{n}$$
$$= S_{xx} - \tfrac{1}{2}\Sigma d_i^2$$

Finally

$$r = \frac{S_{xy}}{\sqrt{S_{xx}S_{yy}}}$$
$$= \frac{S_{xx} - \tfrac{1}{2}\Sigma d_i^2}{\sqrt{S_{xx}^2}}$$

For rank data
$S_{xx} = S_{yy}$

$$= \frac{S_{xx} - \tfrac{1}{2}\Sigma d_i^2}{S_{xx}}$$
$$= 1 - \frac{\tfrac{1}{2}\Sigma d_i^2}{\tfrac{1}{12}n(n^2-1)}$$
$$= 1 - \frac{6\Sigma d_i^2}{n(n^2-1)}$$

Having studied this chapter you should

- be able to use the formula

$$r = 1 - \frac{6\sum d^2}{n(n^2-1)}$$

to calculate the product moment correlation coefficient of data consisting of ranks

- be able to calculate Spearman's coefficient of rank correlation for any set of data points by first ranking of the x values, then ranking the y values and finally calculating the correlation coefficient **of the ranks** using the formula

$$r_s = 1 - \frac{6\sum d^2}{n(n^2-1)}$$

- realise that

  Spearman's coefficient of rank correlation may give a good approximation to the product moment correlation coefficient,

  $r_s = 1$ means the original data points lie on an increasing curve and that the two sets of ranks are identical,

  $r_s = -1$ means the original data points lie on a decreasing curve and that the two sets of ranks are in complete reverse orders.

REVISION EXERCISE

1 The judges gave marks for each of five performers in a talent show. The marks are given in the table below:

Performer	1	2	3	4	5
Mark by judge A	71	32	70	60	52
Mark by judge B	45	83	38	60	54
Mark by judge C	57	40	42	70	39

i) Find Spearman's rank correlation coefficient between the marks given by judge A and the marks given by judge B.

ii) Spearman's rank correlation coefficient between the marks given by judge A and the marks given by judge C is 0.6. State which of the two judges B or C agrees more closely with judge A in ranking the five performers. Give a reason for your answer.

(OCR Jan 2001 S1)

2 In a game show a contestant was asked to identify the years in which each of 10 events occurred. The table below shows the year in which each event actually occurred and the year given by the contestant for that event:

Actual year of occurrence	1963	1968	1971	1973	1983	1984	1986	1990	1991	1997
Year given by contestant	1970	1983	1964	1977	1969	1992	1981	1986	1994	1997

i) Calculate Spearman's rank correlation coefficient between the actual year of occurrence and the year given by the contestant.

ii) Does the value of the correlation coefficient suggest that the contestant is good at remembering the dates of these events? Give a reason for your answer.

(OCR Jun 2001 S1)

3 Two people, Ben and Chandra, were asked to rank three CDs, X, Y and Z, in order of preference. Ben stated that X was best, Y was second best and Z was the least good. Chandra chose her order of preference at random.

i) How many different orders could Chandra choose?

Let R be the value of Spearman's rank correlation coefficient between Chandra's order of preference and Ben's. The tables below show two of the possible orders Chandra could choose, and the resulting value of R.

Case 1

	X	Y	Z
Ben's ranking	1	2	3
Chandra's ranking	1	2	3

The resulting value of R is 1.

Case 2

	X	Y	Z
Ben's ranking	1	2	3
Chandra's ranking	2	1	3

The resulting value of R is 0.5.

ii) Calculate the value of R for each of the other possible orders that Chandra could choose.

iii) Hence show that the distribution of R is given by the table below:

r	-1	-0.5	0.5	1
$P(R = r)$	$\frac{1}{6}$	$\frac{1}{3}$	$\frac{1}{3}$	$\frac{1}{6}$

iv) Find E(R) and Var(R).

(OCR Jan 2003 S1)

4 Two directors of a company each interviewed seven candidates for a job. Director 1 gave each candidate a score from 0 to 100 and a high score meant that the director liked the candidate. Director 2 ranked the candidates and gave the candidate she liked most a rank of 7 and the candidate she liked least a rank of 1. The results are given in the table below:

Candidate	A	B	C	D	E	F	G
Director 1	74	83	64	58	27	91	28
Director 2	6	5	2	4	3	7	1

i) Calculate Spearman's rank correlation coefficient between the assessments given by Director 1 and Director 2.

ii) What can be said about the amount of agreement between the two directors in their assessment of the 7 candidates?

(OCR May 2002 S1)

5 Three trainee business consultants, L, M and N, were required to allocate trustworthiness ratings to seven companies. Their ratings were as follows:

Company	A	B	C	D	E	F	G
L	90	80	70	60	50	40	30
M	48	61	55	93	39	69	82
N	75	70	65	60	80	20	55

Calculate the value of Spearman's rank correlation coefficient between L and M.

The value of Spearman's rank correlation coefficient between L and N is 0.61, and between M and N is −0.82, each correct to 2 significant figures. State, with a reason, which pair of trainees gave ratings which agree most closely with each other.

The value of Spearman's rank correlation coefficient between N's ratings and the ratings of a panel of experts is −1. Obtain the value of Spearman's rank correlation coefficient between M's ratings and the ratings of the panel of experts.

(OCR Jun 2000 S1)

6 The gains made on consecutive Fridays in July 1997 by nine Unit Trusts are given in the following table:

Trust	A	B	C	D	E	F	G	H	I
July 18	+0.25	+0.28	+0.57	+1.62	+1.72	+1.80	+2.27	+2.70	+4.10
July 25	+0.25	−0.15	+0.26	+0.14	+0.20	−0.70	−1.65	−1.50	+1.70

Calculate Spearman's rank correlation coefficient for the data.

(OCR Nov 1999 S1)

7 After a weekly shopping trip, a shopper was asked to write down, from memory, the prices of a selection of seven items purchased. The table shows the actual prices of the seven items and the prices written down by the shopper:

Actual price (p)	77	32	83	45	112	94	56
Price written down (p)	55	35	95	50	121	83	75

i) Calculate the value of Spearman's rank correlation coefficient between the actual price and the price written down.

ii) Does the high value of Spearman's rank correlation coefficient indicate that the shopper is good at remembering prices? Give a reason for your answer.

(OCR Mar 1998 S1)

8 Two students, Arif and Beth, collected the following data relating the mean diversity d of plant species with the distance s meters up an irregular cliff face:

s	0	1.25	2.5	3.85	5.2	6.5	7.8	9.1
d	8.17	8.65	7.47	7.77	6.80	7.21	6.23	6.77

$[n = 8, \sum s = 36.2, \sum d = 59.07, \sum s^2 = 235.575, \sum d^2 = 440.6151, \sum sd = 251.828]$

i) Arif finds the product moment correlation between s and d. Calculate the answer he should get.

ii) Beth finds Spearman's rank correlation coefficient between s and d. Calculate the answer she should get.

iii)

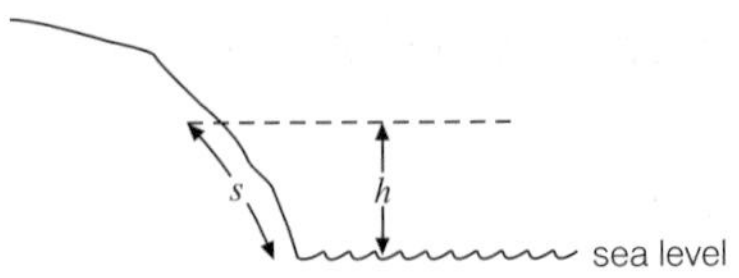

sea level

Subsequently the vertical heights, h metres, above sea level were measured (see diagram). The students now find their respective correlation coefficients between h and d. State, with a reason in each case, whether

a) Arif, or
b) Beth

should obtain the same answer as before.

(OCR Jun 1999 S1)

Representation of data

Revise Chapters 1, 2 and 3 before attempting this exercise.

1 The times t (in seconds) taken by an athlete to run 400 metres on 10 successive days were

53.2, 55.7, 54.2, 52.7, 53.6, 56.8, 54.0, 53.7, 59.3, 53.8

[If required, you may use $\sum t = 547.0$, $\sum t^2 = 29\,957.48$.]

i) Calculate the mean of the times.

ii) Calculate the standard deviation of the times.

iii) Determine the median of the times.

(OCR Nov 1997 P2)

2 Examinations in English, Mathematics and Science were taken by 400 students. Each examination was marked out of 100 and the cumulative frequency graphs illustrating the results are shown below:

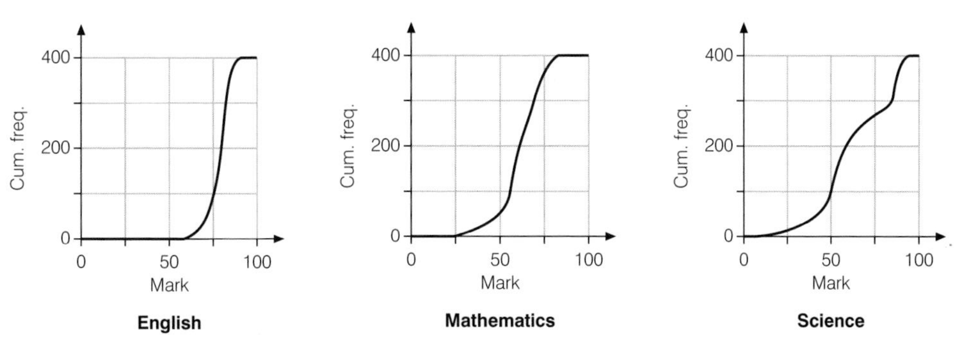

i) In which subject was the median mark the highest?

ii) In which subject was the interquartile range of the marks the greatest?

iii) In which subject did approximately 75% of the students score 50 marks or more?

(OCR Mar 1999 P2)

3 Data were collected from a survey of 200 male students in a college canteen. Their daily expenditure on mid-day meals is summarised in the table below:

Daily expenditure (£)	Frequency
0.50 – 1.49	23
1.50 – 1.74	36
1.75 – 1.99	62
2.00 – 2.49	41
2.50 – 3.49	38

Illustrate the data by means of a cumulative frequency graph. Hence estimate the median and interquartile range of the daily expenditures. Draw a box plot to summarise these calculations.

The diagram below is a box plot for the daily expenditure of a sample of 150 female students at the college canteen.

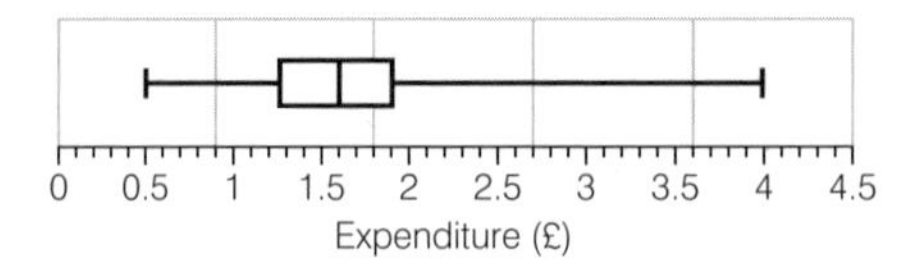

Use this box plot and your earlier work to make two comparisons between the daily expenditure at the canteen of male and female students.

4 At a bird observatory, migrating willow warblers are caught, measured and ringed before being released. The histogram on the right illustrates the lengths, in millimetres, of the willow warblers caught during one migration season.

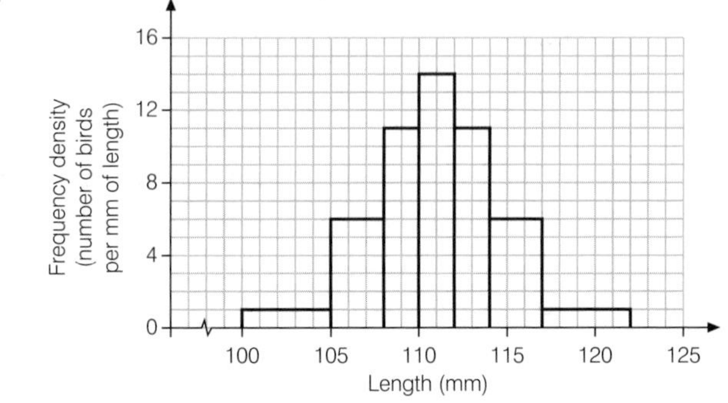

i) Explain how the histogram shows that the total number of willow warblers caught at the observatory during the migration season is 118.

ii) State briefly how it may be deduced from the histogram (without any calculation) that an estimate of the mean length is 111 mm. Explain briefly why this value may not be the true mean length of the willow warblers caught.

iii) Given that the lengths, x mm, of the willow warblers caught during this migration season were such that $\sum x = 13\,099$ and $\sum x^2 = 1\,455\,506$, calculate the standard deviation of the lengths.

(OCR Nov 1998 P2)

5 A sample of 20 observations $x_1, x_2, \ldots x_{20}$ can be summarised by

$$\sum x = 130 \qquad \sum x^2 = 925$$

a) Calculate the mean and standard deviation of these observations.

New values $y_1, y_2, \ldots, y_{20}$ are obtained from $x_1, x_2, \ldots, x_{20}$ using the rule

$y_i = 2x_i - 3.$

b) Obtain the mean and standard deviation of $y_1, y_2, \ldots, y_{20}$.

6 A canoe club held a regatta in which all the members canoed a distance of 2 km. The times, in minutes, for the adult males (aged 18–49) and the senior males (aged 50 and above) are shown in the stem and leaf diagram below:

Adult males		Senior males
9	**1**	6
7 6 3 1	**2**	4 5 8
8 8 5 4 4 3	**3**	2 3 7 7 9 9
9 6 2 0	**4**	3 5
	5	2

Key:
2 | 8 represents a time of 28 minutes

a) Geoff is a senior male member of the canoe club who completed the 2 km in 24 minutes. How many male members of the club completed the course in less time?

b) Find the median and quartiles of the times taken by the adult male members to canoe the distance.

c) Find the median and quartiles of the times taken by the senior male members to canoe the distance.

Use the stem and leaf diagram and your answers to (b) and (c) to compare the times of the senior males with those of the adult males.

7 A bookshop has 200 paperbacks and 300 hardbacks for sale. The paperbacks have prices, $£p_1, £p_2, ..., £p_{200}$ distributed with a mean of £9.50 and a standard deviation of £2.20. The hardbacks have prices $£h_1, £h_2, ..., £h_{300}$ where

$$\Sigma (h_i - 12) = 180 \qquad \Sigma (h_i - 12)^2 = 3996$$

a) Calculate the mean and standard deviation of the prices of the hardback books.

b) Find the value of Σh_i and prove that $\Sigma h_i^2 = 51\ 516$.

c) Find the values of Σp_i and Σp_i^2.

d) Hence find the mean and standard deviation of the prices of the combined sample of 500 books.

8 A group of 20 people went on a trip to the seaside. The total cost, in pounds, of each person's outing is shown below:

67 70 43 34 35   47 83 25 37 58
39 46 66 73 73   56 63 38 60 51

a) Represent this data in a stem and leaf diagram.

b) Obtain the median and interquartile range of this data.

c) Draw a box plot to represent the data.

The total cost of each person's outing consists of a charge of £20 for the hire of the coach used to travel to the seaside plus the money spent whilst they were at the seaside.

d) Without doing any further calculations, write down the median and interquartile range of the money spent by these people whilst they were at the seaside.

9 The diagram shows a cumulative frequency graph of the times spent by the diners at a restaurant one evening:

a) Obtain the median and interquartile range of the data.

b) Produce the frequency distribution for the data and hence draw a histogram for the data.

c) Calculate estimates for the mean and standard deviation of the data.

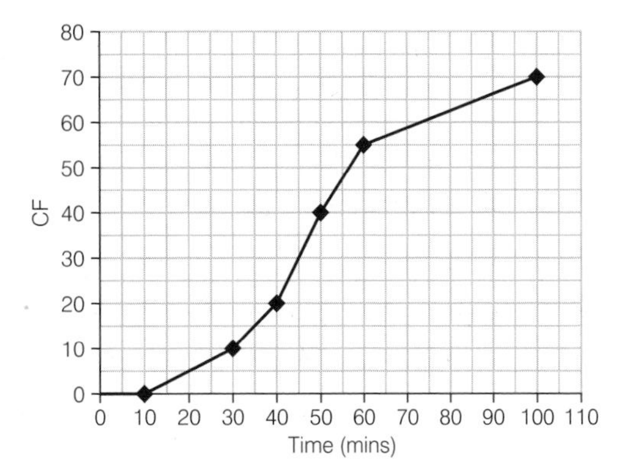

REVISION 2 Probability

Revise Chapters 4, 5 and 6 before attempting this exercise.

1 A committee of four people are to be chosen from a group of 18 men and 12 women.

a) In how many different ways can the committee be selected?

b) How many different committees comprising three men and one woman can be selected?

If the committee is chosen at random find the probability that:

c) the committee consists of three men and one woman;

d) the committee contains at least one woman.

2 A die is biased so that, when rolled, the probability of scoring a 6 is $\frac{1}{4}$. The probabilities of obtaining each of the five other scores 1, 2, 3, 4, 5 are all equal.

a) Calculate the probability of obtaining a score of 5 with this biased die.

b) The biased die and an unbiased die are now rolled together.
Calculate the probability that the total score is 11 or more.

c) The two dice are rolled again. Given that the total score is 11 or more, calculate the probability that the score on the biased die is 6.

(OCR Nov 1995 P1)

3 Two netball teams, A and B, play a series of three matches against each other. The probability that A wins the first match is 0.6. For each subsequent match the probability of a team winning the match is 0.8 if that team won the previous match. There are no drawn matches, so every match results in a win for either A or B.

a) Show this information in a tree diagram.

b) Hence find the probability that:

i) A wins all three matches;

ii) A wins one match and B wins two matches.

4 A bag contains 3 red, 4 blue and 6 green marbles. The marbles are to be taken out of the bag, one at a time, and placed in a line.

a) How many different possible arrangements of the marbles are there?

b) How many arrangements have a green marble at each end of the line?

c) How many of the arrangements have the same colour marble at each end of the line?

d) If each possible arrangement is equally likely, find the probability that the marbles at each end of the line are different colours.

5 A group consisting of five boys and three girls is to be arranged in a line for a photograph.

a) How many different arrangements of the eight people are there?

The photographer suggests that the group should arrange itself in such a way that no two girls are standing next to each other.

b) In how many ways can the group arrange themselves to satisfy the photographer's suggestion?

6 A company makes light meters, and the probability that a randomly chosen meter is faulty is 0.04. In a quality control process, each meter is checked and either accepted or rejected. For a faulty meter, the probability that it will be rejected is 0.84. For a meter with no faults, the probability that it will be rejected is 0.01. Find the probability that:

a) a randomly chosen meter will be faulty and accepted;

b) a randomly chosen meter will be accepted;

c) a randomly chosen meter will be faulty given that the quality control process rejects it.

(OCR Mar 1996 P1)

7 C and D are two archers. With each arrow, C has probability 0.2 of hitting the bulls-eye and D has probability 0.3 of hitting the bulls-eye. They decide to play a game in which each archer starts with two arrows, they take turns, with C starting, to fire arrows at the target and the first person to hit the bulls-eye wins the game. If no-one hits the bulls-eye with one of their two arrows then the game is drawn.

a) Prove that the probability that C wins the game is 0.312.

b) Find the probability that the game is drawn.

c) Find the probability that C wins the game given that the game was not drawn.

8 A group of 12 people, including Mr and Mrs Jones and their son, Oliver, have booked 12 seats, in a row, at the cinema.
In how many ways can the group sit in the seats if:

a) there are no restrictions;

b) Mr and Mrs Jones and Oliver must sit together;

c) Mrs Jones must not sit next to Oliver?

9 Forty 17- and 18-year old students are the only people present at a party. The number of male and female students of each age are given in the following table:

	17-year old	18-year old
Male	9	13
Female	7	11

In the Grand Draw, each of the 40 students has an equal chance of winning one of two prizes. The first prize is a gift token and the second prize is a box of chocolates. No student may win more than one prize. Find the probability that:

a) the gift token is won by an 18-year old male student;

b) both prizes will be won by female students;

c) the box of chocolates will be won by a 17-year old student given that the gift token is won by a 17-year old male student.

(OCR Feb 1997 P1)

10 A coin is biased so that, on each toss, the probability of obtaining a head is 0.3. The coin is tossed three times. By drawing a tree diagram, or otherwise, find the probability that:

a) at least one head is obtained;

b) exactly two heads are obtained;

c) exactly two heads are obtained given that at least one head is obtained.

REVISION 3 Discrete Random Variables

Revise Chapters 7, 8 and 9 before attempting this exercise.

1 The random variable V has the probability distribution given in the following table:

v	1	2	3	4
P(V) = v)	$\frac{1}{3}$	k	$\frac{1}{6}$	k

i) Find the value of the constant k.

ii) Calculate the expectation and the variance of V, giving your answers as exact fractions.

(OCR Jun 1999 S1)

2 When I make a telephone call to an office, the probability of not getting through is 0.45. If I do not get through then I try again later. Let X denote the number of attempts I have to make in order to get through. Starting any necessary assumption, identify the probability distribution of X.

Hence calculate:

i) $P(X \geqslant 4)$;

ii) E[X].

(OCR Mar 1997 S1, adapted)

3 It is given that 33.8% of 20-year olds have achieved at least one A level pass. A sample of 11 20-year olds is selected. Stating one necessary assumption, calculate the probability that, out of the sample of 11,

i) exactly two,

ii) more than three

have achieved at least one A level pass.

(OCR Nov 1999 S1)

4 Peter picks a counter at random from a bag containing 4 red and 6 green counters. Harriet picks a counter at random from a bag containing 6 red and 4 blue counters. The total number of red counters picked by Peter and Harriet is denoted by T.

i) Tabulate the probabilities that T = 0, 1, 2.

ii) Calculate the mean and variance of T.

iii) By considering the mean and variance of the distribution $B(n, p)$, or otherwise, show that the distribution of T is not binomial.

(OCR Mar 2000 S1)

5 On a production line 12% of completed items are faulty. Each completed item is tested. The number of items that have been tested when the first faulty item is found is X. Stating a necessary assumption, suggest an appropriate model for the distribution of X.

Using your model, find:

i) $P(X \geqslant 3)$;

ii) E[X].

(OCR Mar 1996 S1)

6 The random variable X has the distribution B(20, p).

i) Given that $p = 0.7$, find P(X = 18).

ii) Given that $p = 0.7$ and P(X ⩾ x) > 0.1, find the greatest possible value of x.

iii) Given that P(X = 20) = 0.02, find the value of p.

iv) Given that P(X = 10) = 0.02, show that $p(1 - p) = 0.2011$ correct to 4 decimal places, and hence find the two possible values of p.

(OCR Mar 2000 S1)

7 The discrete random variables X takes the values 1, 2, 3, 4 and 5 only, with the probabilities shown in the table:

x	1	2	3	4	5
P(X) = x)	a	0.3	0.1	0.2	b

i) Given that E[X] = 2.34, show that $a = 0.34$ find the value of b.

ii) Find Var[X].

(OCR Mar 1997 S1)

8 A do-it-yourself winerack kit contains 33 rods, which have to be fitted into appropriate sockets. The number of badly fitting rods in a randomly chosen kit is denoted by X. State what needs to be assumed about the rods in a kit in order to model the distribution of X by a a binomial distribution.

Given that the probability of any rod fitting badly is 0.05, use a binomial distribution:

i) to find P(X ⩾ 3);

ii) to obtain the mean and variance of X, giving your answers to 3 significant figures.

(OCR Jun 1996 S1)

9 Before starting to play the game 'Snakes and Ladders' each player throws an ordinary unbiased die until a six is obtained. The number of throws before a player starts is the random variable Y, where Y takes the values 1, 2, 3, ...

i) Name the probability distribution of Y, stating a necessary assumption.

ii) Find E[Y].

iii) Two people play Snakes and Ladders. Calculate the probability that they will each need at least 5 throws before starting.

(OCR Mar 1998 S1)

10 On average 3% of the CDs produced in a factory are defective. CDs are packed in boxes of 50.

a) An inspector chooses 10 CDs at random from a box and tests them. Use a binomial distribution to calculate, correct to 3 decimal places, the probability that the number of defective CDs in this sample is:

i) zero,

ii) exactly one,

iii) two or more.

b) If none of these 10 CDs is defective, no further CDs are tested. If exactly one of the 10 CDs is defective, a further 10 CDs from the box (but no more) are tested. If two or more of the original 10 CDs are defective, all the remaining 40 CDs are tested.

i) Tabulate the total number of CDs in the box that are tested, and the corresponding probabilities.

ii) Hence calculate the expected value, and the variance, of the number of CDs tested.

c) Explain why a binomial distribution might not be an appropriate model for the number of defective CDs in a box.

(OCR Jun 1998 S1)

11 Susan is playing a board game in which she throws a fair four-sided die with faces numbered 1, 2, 3 and 4. At each turn the number of points she scores is calculated as follows.

If she throws a number greater than 1, she scores 10 times the number on the die. If she throws a 1, she must throw the die again and she scores twice the number obtained on this second throw.

Calculate the expectation and the variance of Susan's score.

(OCR Jun 1997 S1)

12 When I send an e-mail to Shunichi there is a probability of 0.85 that I shall receive a reply. If I send 12 e-mails to Shunichi, let X denote the number of replies that I receive.

a) Suggest a suitable probability distribution for X and state one assumption that must be made if this distribution is to be appropriate.

b) Hence calculate:

i) $P(X < 10)$;

ii) the mean and standard deviation of X.

Let Y denote the number of e-mails that I must send to Shunichi in order to receive a reply.

c) Suggest a suitable probability distribution for Y.

d) Hence calculate:

i) $P(Y < 3)$;

ii) the expected value of Y.

Bivariate Data

Revise Chapters 10 and 11 before attempting this exercise.

1

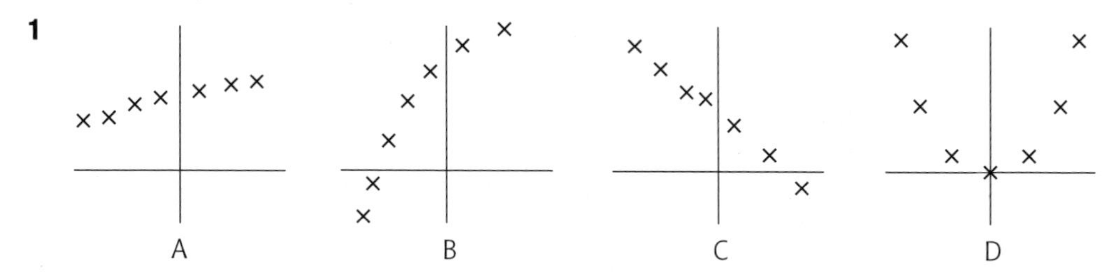

Four sets of biavariate data, A, B, C and D, are displayed in the scatter diagrams. State which of A, B, C and D have product moment correlation coefficients close to 1.

(OCR Nov 1999 S1)

2 Three students, Alex, Beth and Charlie, give scores out of 20 to six student clubs, as follows:

Club	1	2	3	4	5	6
Alex	20	18	14	10	16	7
Beth	11	14	16	17	19	20
Charlie	8	12	11	13	9	10

Calculate the value of Spearman's rank correlation coefficient between Alex's and Beth's rankings.

The value of Spearman's rank correlation coefficient between Beth's and Charlie's rankings is 0.086, correct to 3 decimal places. State what the values of the rank correlation coefficients tell you about the scores given by:

i) Alex and Beth;

ii) Beth and Charlie.

(OCR Mar 2000 S1)

3 Data published by London Research Centre and the Department of Transport give the proportion x% of unemployed adults, and the length y km of roads, in London boroughs in 1991. The data can be summarised as follows:

$n = 33$, $\sum x = 420.2$, $\sum y = 13\ 247$, $\sum x^2 = 6079.82$, $\sum y^2 = 6\ 157\ 699$, $\sum xy = 159\ 149.1$

Calculate the product moment correlation coefficient.

A councillor argues that "This calculation shows that unemployment in our borough can be reduced by building more roads". State, with a reason, whether or not you agree with the logic of this argument.

(OCR Jun 1997 S1)

4 This table gives paired values of the variables g and h:

g	17	23	25	30	32	41
h	5	10	15	20	25	30

$[n = 6, \sum g = 168, \sum h = 105, \sum g^2 = 5048, \sum h^2 = 2275, \sum gh = 3320]$

i) Obtain the equation of the line of regression of g on h. Give your answer in the form $g = p + qh$, stating the values of p and q.

ii) It is required to estimate the value of h when $g = 35$. State when the regression line of g on h, rather than that of h on g, is the correct one to use. Use the regression line of g on h to find the required estimate.

(OCR Jun 1999 S1)

5

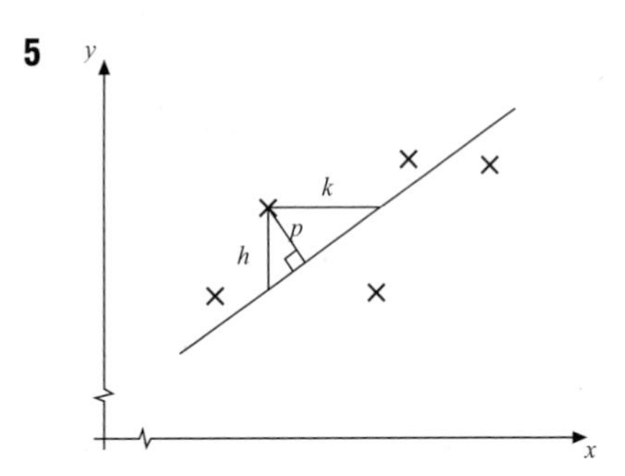

The scatter diagram shows a sample, of size 5, of bivariate data, together with the regression line of x on y. Also shown are the vertical, perpendicular and horizontal distances, h, p and k, respectively, from one of the data points to the line. State which of $\sum h^2$, $\sum p^2$ or $\sum k^2$ is minimised in obtaining the equation of this regression line.

State, giving a reason, whether, for this set of data, the regression line of y on x will be the same as that of x on y.

The data represents the height x and weight y of a randomly chosen sample of men. Which of the two regression lines would you use to estimate the height of a randomly chosen man given his weight?

(OCR Mar 1997 S1)

6 In a ladies' fashion competition, two judges gave marks to seven contestants as follows:

Contestant	*A*	*B*	*C*	*D*	*E*	*F*	*G*
Judge X	64	65	67	78	79	80	96
Judge Y	70	63	61	81	80	78	98

$[\sum x = 529, \sum x^2 = 40\,751, \sum y = 531, \sum y^2 = 41\,239, \sum xy = 40\,948]$

i) Calculate the product moment correlation coefficient.

ii) Calculate Spearman's rank correlation coefficient.

(OCR Jun 1996 S1)

7 An old film is treated with a chemical in order to improve the contrast. Preliminary tests on nine samples drawn from a segment of the film produce the following results:

Sample	*A*	*B*	*C*	*D*	*E*	*F*	*G*	*H*	*I*
X	1.0	1.5	2.0	2.5	3.0	3.5	4.0	4.5	5.0
Y	49	60	66	62	72	64	89	90	96

The quantity x is a measure of the amount of chemical applied, and y is the contrast index, which takes values between 0 (no contrast) and 100 (maximum contrast).

i) Plot a scatter diagram to illustrate the data.

ii) It is subsequently discovered that one of the samples of film was damaged and produced an incorrect result. State which sample you think this was.

In all subsequent calculations this incorrect sample is ignored. The remaining data can be summarised as follows:

$$\sum x = 23.5,\ \sum y = 584,\ \sum x^2 = 83.75,\ \sum y^2 = 44\,622,\ \sum xy = 1883,\ n = 8$$

iii) Calculate the product moment correlation coefficient.

iv) State, with a reason, whether it is sensible to conclude from you answer to part (iii) that x and y are linearly related.

v) The line of regression of y on x has equation $y = a + bx$. Calculate the values of a and b, each correct to 3 significant figures.

vi) Use your regression equation to estimate what the contrast index corresponding to the damaged piece of film would have been if the piece had been undamaged.

vii) State, with a reason, whether it would be sensible to use your regression equation to estimate the contrast index when the quantity of chemical applied to the film is zero.

(OCR Nov 1995 S1)

8 It is known that the wind causes a "chill factor", so that the human body feels the temperature to be lower than the actual temperature. The following table gives the perceived temperature (t °F) for different wind speeds (w miles per hour) when the actual temperature is 25 °F.

w	0	5	10	15	20	25	30	35	40	45	50
t	25	21	9	1	−4	−7	−11	−13	−15	−17	−17

$$[n = 11,\ \sum w = 275,\ \sum w^2 = 9625,\ \sum t = -28,\ \sum t^2 = 2306,\ \sum wt = -3045]$$

i) Calculate the equation of a suitable regression line from which a value of t can be estimated for a given value of w. Simplify your answer as far as possible, giving the constants correct to 3 significant figures.

ii) Use your equation to estimate the perceived temperature when the wind speed is:

a) 38 miles per hour;

b) 55 miles per hour.

iii) Calculate the value of the product moment correlation coefficient for the data, and state what this indicates about the data.

iv) Comment on the reliability of the two estimates found in part (ii).

(OCR Mar 1998 S1)

SAMPLE PAPER

1 The stem-and-leaf diagram shows the heights of some plants, measured correct to the nearest centimetre.

0	5 7 7 8
1	1 2 3 5 6 9
2	0 0 2 4 5 6 7
3	0
4	9

Key: 1 | 2 means 12

i) Find the median and interquartile range of these heights. [3]

ii) State one advantage of using the interquartile range rather than the standard deviation as a measure of the variation in these heights. [1]

iii) State one advantage and one disadvantage of using a stem-and-leaf diagram rather than a box-and-whisker plot to represent data. [2]

2 Some pairs of values of the product moment correlation coefficient (r) and Spearman's rank correlation coefficient (r_s) are given below.

i) $r = 0.9$ and $r_s = 1$ **ii)** $r = 1$ and $r_s = 0.9$ **iii)** $r = 0$ and $r_s = 0$

For each pair, state whether it is possible for both values to be true for a single set of bivariate data. Explain your answers briefly. [3]

3 Starting in January, Jill enters a crossword competition each month. She assumes that in any given month, the probability that she will win is $\frac{1}{10}$.

i) Calculate the probability that

a) her first win is in the fourth month, [3]

b) her first win is before the fourth month, [3]

c) she wins exactly once in the first 4 months. [3]

ii) Calculate the expectation of the number of months up to and including her first win. [2]

4 Seven candidates for a job were each given a mark by three interviewers.

Candidate	Adams	Baker	Hill	Jones	Lee	Paul	Smith
Interviewer A	74	63	60	48	46	40	37
Interviewer B	42	50	63	54	61	68	71
Interviewer C	68	50	60	76	80	65	74

i) Calculate Spearman's rank correlation coefficient, r_s, between the marks for interviewers A and B. [5]

ii) The values of r_s for the other pairs of interviewers, correct to two significant figures, are as follows.

A and C: $r_s = -0.43$ B and C: $r_s = 0.11$

For a second interview, the chairman wishes to use only two of these three interviewers. He wants to include views which differ as widely as possible. Which two interviewers should he use? Give your reason. [2]

5 A bag contains 5 red discs and 4 blue discs. 3 discs are removed at random, without replacement. The number of red discs that are removed is denoted by X.

i) Show that $P(X = 2) = \frac{10}{21}$. [3]

The probability distribution of X is given in the table.

x	0	1	2	3
P(X = x)	$\frac{1}{21}$	$\frac{5}{14}$	$\frac{10}{21}$	$\frac{5}{42}$

ii) Given that $E(X) = \frac{5}{3}$, calculate $Var(X)$. [3]

6 The table shows the age (x years) and the circumference (y cm) of the trunk of each of seven trees of a certain species.

Age (x)	11	21	34	51	13	28	45
Circumference (y)	48.8	105.6	158.4	242.4	64.2	156.6	205.4

$\sum x = 203$, $\sum y = 981.4$, $\sum x^2 = 7297$, $\sum y^2 = 168\,215.48$, $\sum xy = 34\,964.8$]

i) Show that the regression line of y on x has gradient 4.61 (correct to two decimal places) and find its equation. [5]

ii) Estimate the circumference of the trunk of a tree of this species with age

a) 25 years, [1]

b) 100 years. [1]

It is given that the value of the product moment correlation coefficient for the data in the table is 0.99, correct to two decimal places.

iii) Comment on the reliability of your two estimates. [2]

It is suggested that the data could be used to estimate the length of time required for the trunk of a tree of this species to reach a circumference of 200 cm.

iv) State briefly the difference between the method that would be required for this calculation and the method required for the calculations in part (ii). [1]

7 A consumer protection officer found that 15% of packets of a certain brand of cornflakes are underweight. A random sample of 8 packets is chosen from a certain shop. The number of underweight packets in this sample is denoted by X.

i) State an appropriate distribution with which to model X. Give the value(s) of any parameter(s) and state any assumptions required for the model to be valid. [4]

Assume now that your model is valid.

ii) Use tables to find

a) $P(X \geqslant 3)$ [2]

b) $P(X = 4)$ [2]

iii) A random sample of 8 packets is selected on each of 15 days. Find the probability that there are at least 3 underweight packets in fewer than 2 of these 15 samples. [4]

8 A game is played with 9 cards, numbered with the digits from 1 to 9, and coloured grey or white as shown.

1 2 3 4 5 6 7 8 9

The cards are placed in a single line in random order, regardless of colour, to form a 9-digit number.

i) **a)** How many different 9-digit numbers can be formed? [2]

b) Find the probability that no two cards of the same colour are next to each other. [3]

Now the 9 cards are placed in a single line, with all the grey cards in random order on the left and all the white cards in random order on the right.

ii) **a)** How many different 9-digit numbers can be formed? [2]

b) Find the probability that the 4 and the 5 are separated by more than one other digit. [3]

9 A game is played with a fair, six-sided die. The die is thrown once. If it shows an odd number, then this number is the final score. If it shows an even number then the die is thrown again and the final score is the sum of the numbers shown on the two throws. Find the probability that

i) the final score is 5, [3]

ii) the die was thrown twice, given that the final score is 5. [4]

CHAPTER 1

Exercise 1

1 2.51 2 haircuts

2 6 peas 6.23 peas 6 peas

3 5.87 6 6 marks

4 **a)** 1.86 2 spaces
b) 4.15 4 spaces
c) Vertical line diagram and averages show higher number of spaces in car park during August.

5

No. of passengers	Frequency
3	2
4	5
5	1
6	4
7	3
8	4
9	8
10	3
11	3
12	1
13	1
14	1
15	1

8.03 passengers 8 passengers.

6 **a)** 31.7 **b)** 8.575

CHAPTER 1

Exercise 2

1 **a)** 70 vehicles **b)** 105.9 km/hr

2 143 g 320 portions

3 **b)** 42.6 minutes

4 204 minutes

5 **b)** 35.2% **c)** £128 000

6 **a)** 32 **b)** 9 employees 28.1%

7 **a)** 121 **b)** 27.3% **c)** £6270

CHAPTER 1

Exercise 3

1 £18 300
The inhabitants of the second village have generally higher salaries. The average is over £10 000 more in the second village and, from the histograms, we see that only a small proportion of village 2 have income less than £15 000 but a sizeable proportion of village 1 has income less than £15 000.

2 **a)** 59.5–99.5 minutes; midmark = 79.5 minutes
b) 115.1 minutes **c)** 113 minutes

3 **a)** 10–20 years. **b)** 26.5 years 77%

4 **a)** 4.5–14.5; midmark = 9.5
c) 21.1 cm **d)** 24.1 cm
e) Small number of large saplings leads to higher mean than median

5 **a)** 8.7 hours
b)

Time	Midmark	Frequency
0–6	3	12
6–10	8	40
10–15	12.5	20
15–25	20	8

9.575 hours

6 **a)** 59.06 s
b)

Time	Frequency
0–29.5	20
29.5–59.5	8
59.5–89.5	10
89.5–119.5	8
119.5–179.5	2
179.5–239.5	1
239.5–299.5	1

60.2

c) Information has been lost in the grouping process so this second calculation is only an estimate

CHAPTER 1
Revision Exercise

1 37.6 s

Time (s)	Frequency
0–	20
10–	13
20–	15
30–	7
40–	3
60–	2
80–	0
100–200	0

20.1 s
Access times on Sunday morning are much shorter than on Thursday evening

2 10.6 hours 68%

3 14.8 years

4 75.5

5 69 minutes 234 runners

6 2.98 3 goals

7 112.0

8 71 35%

9 **a)** 800 hrs
b)

Time	Frequency
0–500	23
500–700	155
700–900	143
900–1100	120
1100–1500	59

819.7 hours

10 £3300

Expenditure (£)	Frequency
0–2000	26
2000–4000	22
4000–6000	18
6000–10 000	14

£3675

CHAPTER 2
Exercise 1

1 22 12

2 10.5 lb 8 lb

3 46 y 41 y
The ages in the English town are generally lower (median) and with a smaller spread (IQR and range)

4 52 39

CHAPTER 2
Exercise 2

1 **a)** 17 15 **b)** 125 9.5
c) 39 13

2 1.5 1

3 164 cm 14 cm
The pupils in the second class are generally taller (position of LQ, UQ and median) but the spread of heights in the two classes are similar (range and IQR)

4 £25 £13.50
Much higher expenditures on Friday (median and position of quartiles) with much more variability (range and IQR)

5 **a)** 39 s 36.5 s
c) Calls from monthly rental phones are much longer (position of median and quartiles) and show considerably more variability (range and IQR)

CHAPTER 2
Exercise 3

1 2 days 3.28 days^2 1.81 days

2 55 mins 84 mins^2 9.17 mins

3 **a)** Mean = 34.375 sd = 6.22
b) Mean = −2.4 sd = 5.24

4 Mean = 3.3 sd = 1.65

CHAPTER 2

Exercise 4

1

x	1	2	3	4	5	6
f	5	2	6	3	4	5

$\bar{x} = 3.56$ sd = 1.77

2 $\bar{x} = 179.45$ cm sd = 6.96 cm

3 $\bar{x} = 60.5$ years sd = 16.2 years

4 $\bar{x} = 7.21$ mins sd = 5.01 mins

5

School	Mean	sd
A	41.4	12.1
B	40.9	8.98
C	45.5	11.2

School B has the youngest teachers on average and the least spread in the ages; school C has the oldest teachers, on average; school A has the greatest variation in the ages

6 1.65 1.4

7 2.844 0.930
Using the 2 sd rule, most of the distribution lies between 0.98 and 4.7

8 25.58 1.754
Using the 2 sd rule, most of the distribution lies between 22 and 29

9 2.86

10

Data	Mean	Variance
x	1.8	0.96
y	7.2	8.64
z	10	24

$\bar{y} = 3\bar{x} + 2$, $\bar{z} = 5\bar{x} + 1$, Var $y = 3^2 \times$ Var x, Var $z = 5^2 \times$ Var x

CHAPTER 2

Revision Exercise

1 $\bar{x} = 44.6$ mins sd = 9.27 mins

2 Although the marks in the two subjects have a similar average (median), the central 50% of the Maths marks are very tightly packed compared to the central 50% of the English marks (IQR) whilst the overall spread (range) of the Maths marks is more than that in English

3 $\bar{x} = 108$ km sd = 94.9 km
The average distance was greater in 2004 but the spread of distances was less in 2004

4 Median 114 mins IQR = 23 mins
44%

5 Median 8 mins UQ 14 mins

6 **i)** Median = 75 IQR = 13 mins
iv) Male students have generally higher pulse rates

7 $\bar{x} = 1.93$ sd = 0.716
Using the 2 sd rule, most of the distribution lies between 3.37 and 0.49

8 $\bar{x} = 37\,000$ km sd = 13 300 km
c) Mean on simulated trial will probably be higher: less harsh than real driving. The sd will probably be less on the simulated trial: all tyres receiving similar conditions

9 **i)** Median = 41 UQ = 56
LQ = 32
iii) A's games are generally longer than B's (position of median and quartiles)
Similar spreads in each distribution (range and IQR)
iv) No: time of game tells us nothing about who won the game!

10 **i)** Median A = £13; Median B = £26
ii) IQR for A = £21; IQR for B = £26
iii) Checkout A: lower average price suggesting fewer items
iv)

Amount £	0–	10–	20–	40–	60–100
Frequency	10	14	21	9	6

Mean = £30.33

CHAPTER 3
Exercise 1

1 **a)** $\bar{x} = 7.35$ Variance $= 4.15$
sd $= 2.04$
b) $y = 2100x - 4700$
c) $\bar{y} = £10\,700$
Variance of $y = 1.83 \times 10^7$ £^{2}
$\text{sd}_y = £4280$

2 **a)** $\bar{x} = 6.14$ sd $= 1.51$
c) $\bar{y} = 20.7$ $\text{sd}_y = 7.55$

3 **a)** $\bar{x} = 18$ $\text{sd}_x = 5$
b) $\bar{y} = 116$ $\text{sd}_y = 30$

4 **a)** $\bar{x} = 1.2$ $\text{sd}_x = 0.3$
b) $\bar{y} = 35$ $\text{sd}_y = 3$

5 $\bar{x} = 15.5$ $\text{sd}_x = 2$
$\sum(x - 12) = 140$ $\sum(x - 12)^2 = 650$

6 $\bar{x} = 2$ $\text{sd}_x = 0.4$

7 $\bar{x} = \frac{19}{6}$ $\text{sd}_x = 0.5$

CHAPTER 3
Exercise 2

1 $\bar{x} = 173$ minutes

2 **b)** $\bar{x} = 74.3$ years sd $= 6.92$ years
c) $\bar{x} = 74.2$ years sd $= 5.80$ years

3 **a)** $\bar{x} = 52.8$ minutes
$\text{sd}_x = 14.5$ minutes
b) $\sum x = 1056$
c) $\bar{x} = 55.7$ years sd $= 16.6$ years

4 **a)** $\bar{v} = 23.4$ $\text{sd}_v = 4.7$
d) $\bar{x} = 20.7$ thousand pounds
sd $= 4.26$ thousand pounds

5 $\bar{x} = 89.5$ minutes sd $= 9.38$ minutes

6 $\bar{x} = 51.4$ grams
variance $= 476.5$ grams^2
The measure of spread is much larger since the combined sample is really two distinct populations with a marked gap between the two distributions

CHAPTER 3
Revision Exercise

1 **a)** $\bar{x} = 20.4$ $\text{sd}_x = 5.26$
c) $\bar{x} = 20.1$ sd $= 5.12$

2 **i)** $\bar{b} = 23.3$
ii) $\text{sd}_b = 6.99$
iii) $\bar{x} = 22.5$
iv) sd $= 7.56$
v) Boy: 39 marks
Girls did slightly better on average (median)
The range of the girls' scores was lower than the range of the boys' scores but the interquartile ranges are similar

3 **a)** $\bar{x} = 6.53$ items sd $= 1.38$ items
b) 8.53 1.38

4 **a)** $\bar{m} = 461.5$
b) $\text{sd}_m = 2$ grams
c) $\sum(m + 175) = 7638$

5 **a)**

Age	0–1	1–2	2–3	3–5	5–10	10–15
Frequency	8	12	10	16	15	5

b) $\bar{x} = 4.33$ years sd $= 3.33$ years
Since the data is grouped these can only be regarded as estimates
c) Cars in the supermarket are older on average (means) and there is slightly more variation in their ages (sds)
d) $\bar{x} = 4.97$ years sd $= 3.53$ years

6 $\bar{x} = 0.84$ $\text{sd}_x = 0.32$

7 **a)** Median $\approx$ £28 IQR $\approx$ £16
b) $\bar{x} = £30$ sd $= £11.18$
c) $y = 1.45x - 3$
d) **i)** $\bar{y} = €40.5$ $\text{sd}_y = €16.21$
ii) new median $= €37.6$
new IQR $= €23.2$

8 $\bar{x} = £725$ sd $= £221$

CHAPTER 4
Exercise 1

1 234

2 150

3 40

4 120

5 1680

6 1320

7 42 504

8 5 852 925

9 165

10 1140; 6840

CHAPTER 4
Exercise 2

1 1120

2 **a)** 3060 **b)** 480

3 65 780
a) 20 349 **b)** 13 300

4 23 100

5 **a)** 17 7100 **b)** 47 190
c) 38 760

6 **a)** 12 650 **b)** 210 **c)** 351
d) 1820 **e)** 10 830 **f)** 4725
g) 2430 **i)** 5940

7 20 358 520
a) 55 770 **b)** 167 310
c) 669 240

CHAPTER 4
Exercise 3

1 **a)** $\frac{3}{10}$ **b)** $\frac{5}{10}$ **c)** $\frac{8}{10}$

2 **a)** $\frac{1}{8}$ **b)** $\frac{1}{2}$ **c)** $\frac{1}{2}$ **d)** $\frac{7}{8}$

3 **a)** $\frac{1}{9}$ **b)** $\frac{1}{9}$ **c)** $\frac{1}{2}$ **d)** $\frac{1}{2}$
e) $\frac{5}{12}$ **f)** $\frac{5}{18}$

4 **a)** $\frac{5}{48}$ **b)** $\frac{1}{2}$ **c)** $\frac{1}{12}$

5 **a)** 0.3426 **b)** 0.1028 **c)** 0.814

6 **a)** $\frac{1}{22}$ **b)** $\frac{3}{44}$ **c)** $\frac{28}{55}$ **d)** $\frac{3}{11}$

7 **a)** $\frac{1}{55}$ **b)** $\frac{3}{110}$ **c)** $\frac{27}{55}$ **d)** $\frac{9}{22}$
e) $\frac{21}{110}$

8 **a)** 0.003096 **b)** 0.2066
c) 0.04541 **d)** 0.009288
e) 0.04334

9 **a)** 0.4461 **b)** 0.5539
c) 0.3775 **d)** 0.1922
e) 0.412

CHAPTER 4
Revision Exercise

1 **i)** 495 **ii)** 210 **iii)** $\frac{14}{33}$

2 0.369

3 $\frac{7}{48}$ $\frac{5}{16}$ $\frac{3}{16}$

4 270 725
a) 13 182 **b)** 57 798
c) 82 251 **d)** 188 474
e) 28 561

5 **a)** 11 880 **b)** 495

6 **a)** 0.266 **b)** 0.105 **c)** 0.990

7 **a)** 0.255 **b)** 0.218

8 **a)** 210 **b)** 112 **c)** 60

9 **a)** 3003 **b)** 420
c) 2590 **d)** 0.862

10 **a)** 125 970 **b)** 99 768 240

CHAPTER 5
Exercise 1

1 **a)** 0.73 **b)** 0.70

2 **a)** 0.72 **b)** 0.55

3 **a)** 0.210 **b)** 0.689

4 **a)** 0.404 **b)** 0.635
c) 0.993 **d)** 0.145
e) 0.435

5 **a)** 0.4 **b)** 0.471
c) 0.2 **d)** 0.671

6 **a)** $p = 0.25$ **b)** 0.35

CHAPTER 5
Exercise 2

1 **a)** $\frac{5}{12}$; $\frac{3}{12}$; $\frac{2}{12}$; $\frac{2}{3}$ **b)** True

2 **a)** $\frac{2}{12}$; $\frac{7}{12}$; $\frac{2}{12}$; $\frac{2}{7}$ **b)** True

3 **a)** $\frac{2}{12}$; $\frac{5}{12}$; 0, 0 **b)** True

4 **a)** $\frac{6}{36}$; $\frac{6}{36}$; $\frac{2}{36}$; $\frac{2}{6}$ **b)** True

5 **a)** $\frac{3}{36}$; $\frac{6}{36}$; 0; 0 **b)** True

CHAPTER 5
Exercise 3

1 **a)** 0.3263 **b)** 0.1865 **c)** 0.571

2 **a)** 0.525 **b)** 0.7 **c)** 0.75

3 **a)** 0.5865 **b)** 0.04588
c) 0.0782

4 **a)** $\frac{23}{95}$ **b)** $\frac{55}{69}$

5 **a)** $\frac{35}{51}$ **b)** $\frac{3}{7}$

CHAPTER 5
Exercise 4

1 **a)** 0.56 **b)** $\frac{6}{7}$

2 $\frac{49}{153}$

3 **a)** 0.32 **b)** $\frac{9}{16}$

4 $\frac{13}{25}$ $\frac{523}{1125}$

5 0.615 0.610

6 0.014 0.0205 0.683

7 **a)** 0.0294 **b)** 0.107
c) 0.893 **d)** 0.725

Not for mass screening since 72.5% of positive responses are with healthy people causing unnecessary alarm

8 **a)** 0.05217 **b)** 0.0917
c) 0.2435 **d)** 0.8022
e) 0.304

9 **a)** $\frac{3}{7}$ **b)** $\frac{2}{7}$ **c)** $\frac{5}{42}$ **d)** $\frac{5}{12}$

10 **a)** 0.001 **b)** 0.00599
c) 0.00699 **d)** 0.167

11 **a)** 0.294 **b)** 0.466 **c)** 0.940

12 $\frac{8}{21}$

CHAPTER 5
Exercise 5

1 **a)** $\frac{1}{24}$ **b)** $\frac{1}{78}$ **c)** $\frac{3}{8}$

2 **a)** $\frac{8}{15}$ **b)** $\frac{4}{15}$ **c)** $\frac{1}{15}$

3 **a)** $\frac{1}{5}$ **b)** $\frac{4}{15}$

4 **a)** 0.49 **b)** 0.21

5 **a)** $\frac{64}{729}$ **b)** $\frac{19}{27}$ **c)** $\frac{25}{81}$ **d)** $\frac{152}{729}$

6 0.48 0.26 0.26 0.48

7 **a)** $\frac{10}{24}$ **b)** $\frac{9}{10}$

8 **a)** 0.68 **b)** 0.39 **c)** 0.30

9 0.3636 0.515

10 **a)** 0.07 **b)** 0.93
c) 0.38 **d)** 0.81

11 **a)** $\frac{3}{14}$ **b)** $\frac{9}{14}$ **c)** $\frac{1}{32}$ **d)** $\frac{55}{343}$
e) $\frac{54}{343}$ **f)** $\frac{1}{16}$

12 $\frac{22}{45}$ $\frac{322}{855}$ $\frac{22}{45}$ **d)** $\frac{88}{437}$

CHAPTER 5
Revision Exercise

1 **i)** $\frac{1}{6}$ **ii)** $\frac{2}{15}$ **iii)** $\frac{79}{180}$

2 0.108 0.352

3 **a)** $\frac{4}{12}$ **b)** $\frac{5}{11}$ **c)** $\frac{5}{44}$ **d)** $\frac{5}{12}$
e) $\frac{15}{22}$

4 **a)** $\frac{1}{3}$ **b)** $\frac{2}{15}$ **c)** $\frac{7}{30}$ **d)** $\frac{23}{30}$

5 0.228 0.208 0.891
0.1309 0.830

6 $\frac{206}{216}$

7 **a)** $\frac{3}{14}$ **b)** $\frac{15}{28}$ **c)** $\frac{45}{392}$ **d)** $\frac{249}{392}$

8 0.63 0.34
Assume performance in each exam is independent of other exam

9 **a)** **i)** $\frac{4}{15}$ **ii)** $\frac{6}{14}$
b) **i)** 0.145 **ii)** 0.0747

10 **i)** 0.02 **ii)** 0.98
0.184
iii) 0.56 **iv)** 0.0006 **v)** 0.342

CHAPTER 6
Exercise 1

1 15 504

2 3 603 600

3 **a)** 120 **b)** 625

4 **a)** 177 100 **b)** 61 425 **c)** 5215
d) 138 075

5 **a)** 0.0001779 **b)** 0.00213

6 161 280

7 **a)** 21 034 470 600
b) 4 181 076 900

8 **a)** 0.397 **b)** 0.636

9 **a)** 495 **b)** 15 **c)** 450

CHAPTER 6
Exercise 2

1 **a)** 40 320 **b)** 10 080 **c)** 30 240

2 **a)** 720 **b)** 144 **c)** 480
d) 36 **e)** 72

3 50 400

4 1120

5 **a)** 40 320 **b)** 2880 **c)** 0.0714

6 1 663 200

7 362 880
a) $\frac{5}{9}$ **b)** $\frac{4}{9}$ **c)** 0.0476 **d)** $\frac{7}{9}$

8 **a)** 144 **b)** 1440

9 **a)** 10 368 000 **b)** 933 120
c) 4 790 016 000 **d)** 72 576 000

10 2880

11 **a)** 70 **b)** 112 **c)** 28
210

CHAPTER 6
Revision Exercise

1 **i)** 720 **ii)** 240 **iii)** $\frac{1}{3}$

2 4725

3 **a)** 4845 **b)** 1760 **c)** 0.465

4 **a)** 3 628 800 **b)** 725 760
c) 34 560 **d)** 604 800

5 **a)** 27 720 **b)** 4200
c) 0.288

6 **a)** 12 600 **b)** 1680 **c)** 2940

7 **i)** 40 320 **ii)** 1152 **iii)** 53

8 **i)** 0.198 **ii)** 0.984 **iii)** 1260

9 **i)** 2520 **ii)** 168
iii) 0.311

10 **a)** 420 **b)** $\frac{13}{35}$

CHAPTER 7
Exercise 1

1

x	3	4	5
y	0.28	0.3744	0.3456

2 $k = \frac{1}{32}$

x	2	3	4	5
y	$\frac{5}{32}$	$\frac{7}{32}$	$\frac{9}{32}$	$\frac{11}{32}$

$\frac{12}{32}$

3 $q = 0.12$
b) 0.41
c) 0.51
d) 4

4 **b)**

r	0	1	2	3
p	$\frac{1}{84}$	$\frac{18}{84}$	$\frac{45}{84}$	$\frac{20}{84}$

c) $\frac{3}{4}$

5

n	1	2	4	8	16
p	$\frac{9}{36}$	$\frac{12}{36}$	$\frac{10}{36}$	$\frac{4}{36}$	$\frac{1}{36}$

2 $\frac{13}{18}$

6

j	0	1	2	3
p	$\frac{1}{30}$	$\frac{9}{30}$	$\frac{15}{30}$	$\frac{5}{30}$

$\frac{1}{3}$

7 **a)** 0.12
b) 0.40
c) 0.14

CHAPTER 7
Exercise 2

1 $4\frac{1}{3}$ 3.54

2 **a)** $q = 0.3$
b) E[X] = 2.1 sd = 1.14

3 **c)** E[X] = 2.875
d) sd = 1.05

4 **a)** $\frac{24}{210}$
b) E[X] = 2.4 Var W = 0.64
c) $\frac{19}{42}$

5

x	1	2	3	4	5	6	7	8	9	10	11	12
p	$\frac{1}{6}$	$\frac{1}{6}$	$\frac{1}{6}$	$\frac{1}{6}$	$\frac{1}{6}$	0	$\frac{1}{36}$	$\frac{1}{36}$	$\frac{1}{36}$	$\frac{1}{36}$	$\frac{1}{36}$	$\frac{1}{36}$

E[X] = $4\frac{1}{12}$ sd = 2.83

6 **a)** $t = 0.3$ $s = 0.1$ sd = 1.56

7 **b)** E[X] = 2.208 Var X = 0.917

8 **b)** $\mu = 4.07$ $\sigma^2 = 0.621$

9 **a)** $p = 0.05$
b) $\mu = 2.95$ $\sigma = 3.06$
c) 0.2

10 **b)** $\mu = 1$ $\sigma = 1.18$
c) 0.5

CHAPTER 7
Revision Exercise

1 **a)** $r = 0.1$, $s = 0.3$
b) $\sigma = 1.47$

2 E[X] = 5.25 $\sigma = 3.22$ $\frac{5}{12}$

3 $\mu = 3$ $\sigma = 0.775$

4 $k = \frac{1}{15}$ E[R] = $2\frac{2}{3}$ $\sigma^2 = \frac{14}{9}$ $\frac{3}{5}$

5 **a)** $k = \frac{1}{5}$
b) E[V] = 3.04 $\sigma^2 = 0.9984$

6 E[X] = 2 $\sigma^2 = 1.2$

7 **a)** 0.6
b) $a = 0.35$, $b = 0.05$
c) $\sigma^2 = 2.5475$

8 $\mu = 2.8$ $\sigma = 0.748$

9 **i)** $a = \frac{1}{2}$, $b = \frac{2}{3}$, $c = \frac{1}{3}$
ii) $\frac{3}{8}$
iii) 1.3125 0.340

10 $p = 0.2$ $\mu = 4.3$ $\sigma^2 = 2.01$

CHAPTER 8
Exercise 1

1 0.1641

2 0.2140

3 0.2837

4 0.3780

5 0.3486
Assumes independence which may well be unlikely at a party.
Constant prob also unlikely — peer pressure, etc.

6 Y is B(20, 0.32)
Constant probability of bus being late (weather etc may mean it varies) and independence of each day (road works on route make this unlikely)
0.1289

7 **a)** $4p^3(1-p)$ **b)** $15p^4(1-p)^2$
c) $p = 0.8416$ **d)** 0.284

8 0.2012

CHAPTER 8
Exercise 2

1 **a)** 0.2188 **b)** 0.2726 **c)** 0.8906

2 **a)** 0.2613 **b)** 0.8414

3 **a)** 0.8338 **b)** 0.2150 **c)** 0.8327

4 **a)** 0.0781 **b)** 0.0031 **c)** 0.9996

5 **a)** 0.2907 **b)** 0.247

6 **a)** 0.6083
b) 0.6741

Part (b) is not binomial since prob of success varies at each selection

7 at least 10 tosses

8 **a)** 0.4992 **b)** $p = \frac{9}{13}$

9 $p = \frac{1}{9}$
0.2869
at least 26 throws

10 $p = 0.1304$
0.8867
0.148

11 $p = 0.0764$

CHAPTER 8

Exercise 3

1 **a)** 20 3.16 **b)** 15 3.54
c) 56 6.35

2 **a)** **i)** $n = 28$ $p = 0.12$
ii) Constant prob of lateness (weather may alter probs) independence of arrivals (one late arrival may affect later times)
b) **i)** 0.6695 **ii)** E[W] = 3.36
iii) $\sigma = 1.72$

3 **a)** $p = 0.8$
$n = 15$
b) 0.6482

4 **a)** 0.4
b) **i)** Ties are replaced each day
ii) 6
1.55
iii) 0.1662

5 **a)** $p = 0.8$
$n = 20$
b) 0.7618

6 **a)** Constant prob of space being occupied (may depend on arrival time, etc.)
Independence of one day to answer
T is B(20, 0.35)
b) E[T] = 7
$\sigma = 2.13$
c) 0.0196

CHAPTER 8

Revision Exercise

1 **a)** 5 is B(14, 0.65)
b) Constant prob of success at first attempt each day
each day's calls independent of other days
c) 0.2178 **d)** 0.9161

2 **a)** Y is B(18, 0.05) **b)** 0.1683
c) **i)** 0.2265 **ii)** $P(Y \leqslant 3) = 0.9891$

3 **i)** 0.5634 **ii)** 0.6257 **iii)** 0.2629

4 **a)** X is B(16, 0.9)
b) Constant prob for each train and each train's punctuality being independent of other trains
c) 0.9316
0.8114
d) $k = 12$ **e)** = 0.5147

5 **a)** 0.2880 **b)** 0.2627 **c)** 0.310

6 **a)** $p = 0.6$ **b)** 0.2173

7 **a)** $n = 20$, $p = 0.85$
b) Constant prob for each (i.e. practice doesn't improve probabilities)
Independence of each shot (i.e. success/failure doesn't affect mental approach next time)
c) 0.4049
d) Y is B(5, 0.4049)
Independence of each day (performance not affected by tiredness, etc.)
e) 0.346
f) 2.02
1.10

8 **a)** ${}_{20}C_5 p^5(1-p)^{15}$
b) ${}_{20}C_4 p^4(1-p)^{16}$
c) 0.385 **d)** 0.187
e) 0.188

9 **a)** 0.499968 **b)** $\frac{1}{243}$ **c)** $p = \frac{12}{13}$

10 **i)** Constant prob of being able to park in space (e.g. arrives at same time each day)
Independence of each day from others (no skip parked there long term!)
ii) **a)** $n = 5$ $p = \frac{2}{5}$ **iii)** 0.882

CHAPTER 9

Exercise 1

1 **a)** 0.125 **b)** 0.9375 **c)** 0.484

2 **a)** 0.0964 **b)** 0.162
c) 0.721 **d)** 0.640

3 **a)** 0.105 **b)** 0.178
c) 0.944 **d)** 0.462

4 X is Geo($\frac{1}{20}$)
Assume constant probability of catching a fish and independence of each cast — may be unlikely one cast may scare away fish
E[X] = 20
0.0429
0.540

5 X is B(20, 0.15)
Independence of each shot; constant prob of bulls-eye — may be unlikely — windy conditions would affect both!
0.6477
Y is Geo(0.15)
Same assumptions
= 0.377

6 **a)** Assuming equal numbers of each card,
X is Geo($\frac{1}{5}$)
E[X] = 5
b) 0.1024 **c)** 0.738 **d)** 0.134

7 **a)** U is Geo(0.17)
Constant prob of exceeding 55 m
Independence of each throw
Weather, tiredness, etc. may make these unlikely
b) 0.0972
0.327
0.428
0.319
c) V is B(6, 0.17)
0.206

8 **a)** $p = 0.4$ **b)** 0.081 **c)** 0.311

CHAPTER 9

Revision Exercise

1 **a)** Constant prob of success for each shot and independence of each shot — unlikely shooting against different goal keepers; dispondency, etc.!
b) **i)** 0.125
ii) 0.185
iii) 0.651

2 **a)** 0.273 **b)** 0.25 **c)** 0.271

3 **a)** T is Geo($\frac{1}{8}$), E[T] = 8 **b)** 0.0957

4 **a)** A is B(3, $\frac{1}{6}$) **b)** B is Geo($\frac{1}{2}$)
c) 0.0694 **d)** $\frac{1}{4}$
e) 0.192

5 **a)** W is B(104, 0.02) **b)** 0.273
c) E[W] = 2.08
$\sigma = 1.43$
d) 0.655 **e)** H is Geo(0.02)
f) 0.350
g) 50

6 **a)** 0.111 **b)** 17.5 **c)** 0.095

7 $p = \frac{1}{9}$
0.0780
0.562

8 T is Geo($\frac{1}{11}$)
Assuming independence + constant probability — depends on real random sample: family groups may be more/less prone to left handedness
i) 0.0683
ii) 0.386
iii) E[T] = 11

9 **a)** 0.101 **b)** 3.125

10 **a)** **i)** 0.0189
ii) 1.429
b) **i)** $(1-p)^4$ **ii)** $p = 0.4$

11 **i)** 0.0462
ii) 0.714
iii) 18.3

CHAPTER 10

Exercise 1

1 $r = -0.885$
Data points are fairly close to a line of negative gradient, r close to -1

2 $r = -0.978$
Data closely packed about line of positive gradient

3 $r = -0.916$
Data reasonably closely packed about line of negative gradient

4 $r = -0.957$
Data reasonably closely packed about line of positive gradient

5 $r = -0.802$
Data reasonably closely packed about line of negative gradient

6 $r = 0.818$
Data reasonably closely packed about line of positive gradient

7 $r = 0.985$

8 $r = -0.799$

9 $r = -0.978$
Data closely packed about line of negative gradient

10 $r = 0.996$
Data closely packed about line of positive gradient

11 $r = 0$
No linear correlation **but** there clearly is a rule linking y and x

12 $r = 0.0207$
No linear correlation

13 $r = 0.995$
Very high positive correlation coefficient due to data consisting of two separated populations

14 For (x, y) data $r = 0.995$
For (u, v) data $r = 0.995$
The coding has no effect on the value of the correlation coefficient

CHAPTER 10

Exercise 2

1 $y = -0.3674x + 210.692$
b) i) $y = 108$
ii) $y = -28$
c) estimate i) is reliable since r is reasonably close to -1 and it is interpolation; estimate ii) is clearly unreliable since £650 000 is well outside the x values given in the data (extrapolation)

2 $y = 0.1482x + 14.128$
b) $y = 27.5$
c) estimate is reliable since r is very close to 1 and x value is within interval of data x values (interpolation)

3 $y = 6.6255x + 7.578$ $y = 54$
estimate is reliable since r is reasonably close to 1 and x value is within interval of data x values (interpolation)
$x = 10.5$ is well outside the interval of data x values (extrapolation) so the estimate would be unreliable

4 $r = 0.9895$ $y = 1.225x + 2.559$
c) $y = 8.7$ $y = 17.3$
d) The x data values range from 0 to 7 so the first estimate is reliable since r is very close to 1 and x value is within interval of data x values (interpolation). The second estimate is an example of extrapolation so should be treated with caution

5 $y = 0.2912x - 0.076$
c) $y = 0.44$

CHAPTER 10

Exercise 3

1 $r = 0.8538$
$x = 0.6929y + 2.535$
Point of intersection is $(\bar{x}, \bar{y})$ or (6, 5)

2 $r = 0.9569$ $x = 6.1799y - 75.935$
$x = 233$ days
Estimate is reliable since r is quite close to 1 and $y = 50$ is within range of y data values (interpolation)

CHAPTER 10

Exercise 4

1 $r = -0.9965$ $y = -0.12x + 10.3$
a) Since r is very close to -1 there is a good linear fit
b) Since the x data is controlled, we find the equation of the y on x line.
This is $y = -0.12x + 10.3$
c) i) $y = 9.22$
ii) $x = 12.5$
iii) $y = 7.3$
d) i) and ii) are reliable since interpolation with r very close to -1; iii) is unreliable since it is an example of extrapolation

2 $y = 0.4856x - 2.397$
$x = 2.0489y + 4.971$
c) $y = 2.46$ (using y on x line)
d) $x = 15.2$ (using x on y line)

3 $y = 0.291x + 4.76$
The load appears to be controlled so the y on x line should be used:
$x = 14.6$

4 $r = 0.9322$ $t = 6.6006n + 50.846$
Since n appears to be controlled and the fact that we want a value of t, we should use the t on n line
$t = 216$ s

5 Since x is controlled, use y on x line for all calculations:
$y = -0.0385x + 25.962$ $y = 16.7$
240 is outside the range of observed x values so the estimate may be unreliable (extrapolation)

6 $r = 0.9843$ $y = 1.92x - 39.805$
ii) $r = 0.9843$, which is close to 1 so the data is close to a straight line of positive gradient
iii) $y = -39.8 + 1.92x$
iv) r is very close to 1 so the two lines will be very close to each other departure time = 0747
v) 120 is well outside the given data values of x: extrapolation is not reliable

CHAPTER 10

Revision Exercise

1 $y = 0.586x + 1.777$
ii) Point of intersection is $(\bar{x}, \bar{y})$ or (7.89, 6.4)
iii) $x = 4.71$

2 $r = 0.9749$
Although r is quite close to 1, it does look as if a non-linear relationship would describe the data

3 $r = 0.9987$ $y = 0.3571x + 0.012$
i) $r = 0.9987$
ii) Use y on x regression to calculate a y value: $y = 0.87$
iii) a) reliable since interpolation and r close to 1
b) unreliable: extrapolation

4 $r = 0.535$ $y = 0.044x - 0.038$
i) $r = 0.535$
iii) $y = 0.072$
iv) Unreliable: extrapolation and modest value of r.

5 $y = 1.395x + 1.703$
i) $r = 0.987$
ii) No change in the value of r since linear coding does not affect correlation coefficient values
iii) Use y on x line since x data is controlled
$x = 70.5$

6 $r = 0.760$ is diagram 3: mild positive correlation
$r = -0.101$ is diagram 1: no clear linear correlation
$r = -0.936$ is diagram 2

7 **a)** $r = 0.971$
b) $d_1^2 + d_2^2 + d_3^2 = 29$
c) $y = 5.25x - 13.25$
d) $d_1^2 + d_2^2 + d_3^2 = 13.5 < 29$

8 **a)** $r = 0.537$
b) $x = 2.907 + 0.6884y$
c) $x = 17.0$
d) Low value of r makes this estimate unreliable

9 $r = -0.2425$
Value of r is quite close to 0 so unlikely to be a linear relationship

10 **i)** $y = 92.155 + 0.102x$
ii) Strictly the x on y line should be used but since r is close to 1 the two lines will be very close so the y on x is acceptable; $y = 140$ is within the range of y data values so interpolation is being used which is reliable

11 **ii)** 1 since points form a straight line of positive gradient
iii) $y = 4.2x - 88$
iv) $y = 100$ $y = 290$
First value is more reliable since $r = 1$ for this part of data whilst on second part of data the straight line fit is not perfect.

12 ii) Use x on y line
$x = 0.477y + 24.2$
$x = 40.9$

ii) $r = 0.861$

iii) a) A reasonable straight line of possible gradient fits the data

b) For middle of class no real relationship evident between 2 marks

CHAPTER 11
Exercise 1

1 a) $r = 0.7714$ **b)** $r = -0.4857$
c) $r = -0.6$
A and B show the greatest agreement (r closest to 1)

2 $r = -0.2619$
No clear agreement between the two judges

CHAPTER 11
Exercise 2

1 $r = 0.8$

2 $r = 0.47619$
Small amount of positive correlation

CHAPTER 11
Exercise 3

1 a) $r = 0.8459$
$r_s = 0.833333$

2 $r_s = 0.939394$

3 a) $r = 0.7994$
b) $r_s = -0.33333$
In this case r_s is a poor estimate of r so the statement is not always true!

4 $r_s = 0.563636$
There is modest positive correlation in the ranked data: probably not enough to feel that the statement is fully justified

5 a) $r = -0.115$
$r_s = 0.257143$
r gives a good indication of whether there is a linear rule linking the data values; r_s is quick to calculate and indicates whether the ranks of the data are in close agreement or not

6 a) Points lie on line of positive gradient
b) Points lie on an increasing curve
c) Points lie on a line of negative gradient
d) Impossible: the points must lie on a line of positive gradient for $r = 1$ and this means that the ranks must be in perfect agreement so $r_s = 1$

CHAPTER 11
Revision Exercise

1 $r_s = -0.8$
The closer r_s is to 1, the better the agreement so A and C are in better agreement than A and B

2 $r_s = 0.745455$
No: the fact that r_s is close to 1 means the contestant is good at getting the dates in the right order but this tells us nothing about the ability to get the exact date correct

3 i) 6 different orders
E[R] = 0; Var[R] = 0.5

4 $r_s = 0.785714$
There is a reasonable level of agreement between the two directors since r_s is quite close to 1

5 $r_s = -0.42857$
L and N are in greatest agreement since their value of r_s is closest to 1
$r_s = 0.82$

6 $r_s = -0.23333$

7 $r_s = 0.928571$
The high value of r_s shows that the shopper is good at putting the prices of items into the right order but does not show that he is good at remembering the actual price since this information is lost in the ranking process

8 i) $r = -0.8646$
ii) $r_s = -0.90476$
iii) Arif would get a different answer since the relationship linking s to h is not linear.
Beth would get the same answer since the ranks for h would be the same as the ranks for d

REVISION 1

Representation of Data

1 **i)** 54.7 s **ii)** 1.91 s **iii)** 53.9 s

2 **i)** English **ii)** Science **iii)** Science

3 Median = £1.92
IQR = 0.66
In general the male students spend more than the female (position of median and quartiles). The variability of the middle 50% of the females expenditure (IQR) is similar to the figure for the boys

4 Area under graph = 118
Graph is symmetrical about $x = 111$
The ungrouped distribution of heights may not be symmetrical about 111: the grouping process loses information
$\bar{x} = 111.01$ cm sd = 3.45 mm

5 **a)** $\bar{x} = 6.5$ sd = 2
b) $\bar{y} = 10$ $\text{sd}_y = 4$

6 **a)** 4 male members did race in under 24 minutes
b) Median = 34 mins
LQ = 26 mins UQ = 40 mins
c) Median = 43 mins
LQ = 33.5 mins UQ = 49 mins
d) Senior males generally took longer (position of medians and quartiles) but spread (IQR) of two sets of data is quite similar

7 **a)** $\bar{h} = £12.6$ $\text{sd}_h = £3.6$
b) $\sum h = 3780$
c) $\sum p = 1900$ $\sum p^2 = 19\,018$
d) $\bar{b} = £11.36$
$\text{sd}_b = £3.47$

8 Median = £53.5 IQR = £28.50
Median = £33.50 IQR = £28.50

9 **a)** Median = 47.5 mins
IQR = 20.8 mins
b)

Time	10–30	30–40	40–50	50–60	60–100
Frequency	10	10	20	15	15

c) Mean = 49.6 mins sd = 19.1 mins

REVISION 2

Probability

1 **a)** 27 405
b) 9792
c) 0.357
d) 0.888

2 **a)** $p = \frac{3}{20}$
b) $\frac{13}{120}$
c) $\frac{10}{13}$

3 0.384 0.176

4 **a)** 60 060
b) 11 550
c) 2310 18 480
d) 0.692

5 **a)** 40 320
b) 14 400

6 **a)** 0.0064
b) 0.9568
c) $\frac{7}{9}$

7 0.312 0.3136 0.455

8 **a)** 479 001 600
b) 21 772 800
c) 399 168 000

9 **a)** $\frac{13}{40}$
b) 0.196
c) $\frac{15}{39}$

10 0.657 0.189 0.288

REVISION 3

Discrete Random Variables

1 $k = \frac{1}{4}$ $E[V] = \frac{7}{3}$ $\text{Var } V = \frac{25}{18}$

2 Independence of each attempt; X is Geo(0.55)
i) 0.0911 **ii)** 1.82

3 Assuming a random sample (not all chosen from a particularly good, or bad, school)
i) 0.153 **ii)** 0.541

4

t	0	1	2
p	0.24	0.52	0.24

$E[T] = 1$ $\text{Var } T = 0.48$

5 X is Geo(0.12) provided items tested are independent of each other.
i) 0.774 **ii)** 8.33

6 **i)** 0.0278
ii) $x_{max} = 17$
iii) 0.822
iv) 0.721, 0.279

7 **a)** $a = 0.34$ $b = 0.06$
b) 1.6644

8 Constant prob of rod fitting badly and each rod being independent of others
i) 0.227
ii) E[X] = 1.65, Var X = 1.5675

9 **i)** Y is Geo($\frac{1}{6}$) if the player doesn't cheat!
ii) E[Y] = 6 **iii)** 0.233

10 **a** **i)** 0.737
ii) 0.228
iii) 0.035

b)

y	10	20	50
p	0.737	0.228	0.035

E[Y] = 13.68 Var Y = 65.3
c) May not be independent – if all CDs come from single machine which is faulty

11 E[S] = 23.75 Var S = 168.4

12 **a)** X is B(12, 0.85)
Assuming independence and constant probability of replies: he may get tired of replying to e-mails!
b) **i)** 0.264
ii) $\mu = 10.2$ $\sigma = 1.24$
c) Y is Geo(0.85)
d) **i)** 0.9775 **ii)** 1.176

REVISION 4

Bivariate Data

1 Graph A

2 $r_s = -0.82857$
A and B are almost in full disagreement
There is little relationship between B and C's rankings

3 $r = -0.385$
This is an example of "nonsense" correlation between two variables with no immediate link. Furthermore value of r is not close to -1

4 **i)** $g = 12.8 + 0.869h$
ii) The g on h line should be used to calculate a value of h if h is a controlled variable
$h = 25.5$

5 The x on y line minimises $\sum k^2$
The two lines will not be the same since $r \neq \pm 1$
x on y line should be used

6 **i)** $r = 0.9515$ **ii)** $r_s = 0.714286$

7 **ii)** Sample F **iii)** $r = 0.9787$
iv) r is very close to 1 so it is reasonable to look for a linear relationship
v) $y = 39.6 + 11.4x$ **vi)** $y = 79.5$
vii) No: 0 is outside the range of data x values and extrapolation is not reliable

8 **i)** $t = -0.853w + 18.8$
ii) $t = -13.6$
$t = -28.1$
iii) $r = -0.946$
Data is close to a line of negative gradient
iv) First result is reliable since it is an instance of interpolation and r is close to -1
Second is unreliable since it is an example of extrapolation

Sample exam paper

1 **i)** 19, 14
ii) Less affected by the 49 which is an outlier
iii) eg: S & L shows original data. Harder to read median from S & L than B & W.

2 **i)** Yes, y always increasing but not linearly
ii) No. $r = 1 \Rightarrow$ linear, hence $r_s = 1$
iii) Yes. eg four points in a square.

3 **i)** **a)** 0.0729 **b)** 0.271 **c)** 0.292
ii) 10

4 **a)** −0.89
b) A and B because the views of this pair are closest to being opposite.

5 **ii)** $\frac{5}{9}$

6 **i)** $y = 4.61x + 6.43$
ii) **a)** 122 **b)** 468
iii) **a)** Reliable because r close to 1.
b) Unreliable because extrapolated.
iv) Use reg. line of x on y instead of y on x.

7 **i)** B(8, 0.15). Proportion underweight in one box is constant and is independent of other boxes.
ii) **a)** 0.105 **b)** 0.0185
iii) 0.523

8 **a)** **i)** 362 880 **b)** $\frac{1}{126}$
ii) **a)** 2880 **b)** $\frac{17}{20}$

9 **i)** $\frac{2}{9}$ **ii)** $\frac{1}{4}$

INDEX